D0153849

EVOLUTION'S BITE

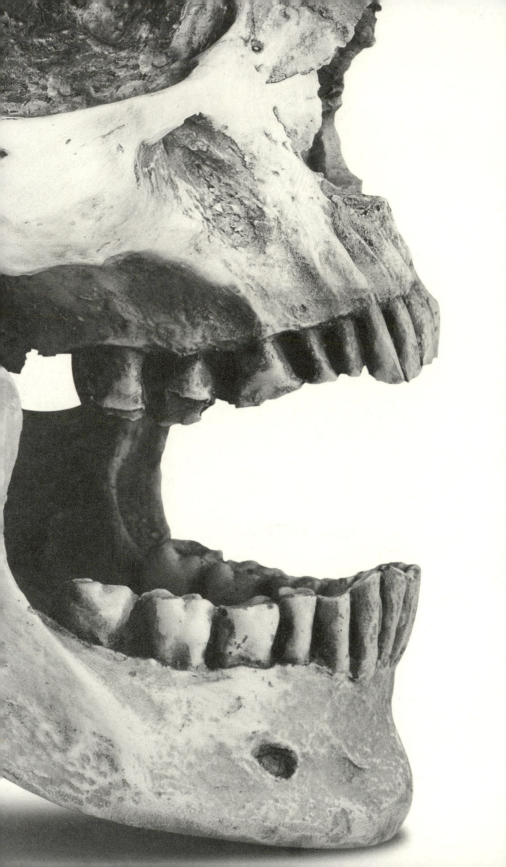

Evolution's Bite

A STORY OF TEETH, DIET, AND HUMAN ORIGINS

PETER S. UNGAR

PRINCETON UNIVERSITY PRESS
PRINCETON AND OXFORD

Copyright © 2017 by Princeton University Press

Published by Princeton University Press,
41 William Street, Princeton, New Jersey 08540

In the United Kingdom: Princeton University Press,
6 Oxford Street, Woodstock, Oxfordshire OX20 1TR

press.princeton.edu

Jacket art courtesy of Shutterstock

All Rights Reserved

ISBN 978-0-691-16053-5

Library of Congress Control Number: 2016960239

British Library Cataloging-in-Publication Data is available

This book has been composed in Garamond Premier Pro

Printed on acid-free paper. ∞

Printed in the United States of America

1 3 5 7 9 10 8 6 4 2

For DIANE, MAYA, & RACHEL

CONTENTS

ACKNOWLEDGMENTS

Much of what I have learned from my mentors, colleagues, and students has found its way into this book. Thank you all. I am grateful to those who read bits and pieces of text, and especially to those generous enough to share their stories. These include Leslie Aiello, Henry Bunn, Thure Cerling, Andrew Cohen, Alyssa Crittenden, Fuzz Crompton, Raymond Dart, Peter deMenocal, Dean Falk, Ann Gibbons, Ken Glander, Fred Grine, Kristen Hawkes, Lyman Jellema, Cliff Jolly, Rich Kay, Joanna Lambert, Leo Laporte, Clark Larsen, Julia Lee-Thorp, Daniel Levine, Peter Lucas, Mark Maslin, Scott McGraw, Andrew Moore, Dani Nadel, Mike Plavcan, Rick Potts, Melissa Remis, Mike Richards, Philip Rightmire, Jerry Rose, Pat Shipman, Becky Sigmon, Matt Sponheimer, Christine Steininger, Bob Sussman, Mark Teaford, Francis Thackeray, Phillip Tobias, Elisabeth Vrba, Alan Walker, Malcolm Williamson, Richard Wrangham, and Barth Wright. I also thank Richard Alley, Henry Bunn, Thure Cerling, Jennifer Clark, Peter deMenocal, John Fleagle, Fred Grine, Alain Houle, Julia Lee-Thorp, Scott McGraw, Andrew Moore, Dani Nadel, Dolores Piperno, Rick Potts, Jerry Rose, Matt Sponheimer, Bob Sussman, and James Zachos for their kind permission to use figures or data used in generating imagery for this book. I thank the two anonymous reviewers for their thoughtful comments and suggestions, Susan Ungar, for help with proof reading, Maia Vaswani for her excellent work copyediting this book, and Alison Kalett, Jenny Wolkowicki, and the rest of the staff of Princeton University Press for their support and encouragement throughout the process.

EVOLUTION'S BITE

INTRODUCTION

We hold in our mouths the legacy of our evolution. Nature has sculpted our teeth over countless generations into tools adapted to chew the foods our ancestors had to eat to survive. And paleoanthropologists, those of us who study human origins, spend a lot of time thinking about them. Teeth are our bridge to the past. They allow us to track changes from one species to the next to trace our evolution. Yes, we have fossilized skulls and skeletons to work with too, but teeth are special. They are essentially ready-made fossils that have remained virtually unchanged for millions of years. More important, they are the most commonly preserved part of the digestive system, and the key to unlocking the diets of our ancestors. We look to tooth size, shape, pattern of wear, and chemistry to work out details of the foods eaten by long-gone species.

Because we can use teeth to reconstruct diet, they are also the key to unlocking an extinct species' place in nature. In *Love and Death*, Woody Allen's character Boris Grushenko described nature as "big fish eating little fish, and plants eating plants, and animals eating an . . ." He continued, "It's like an enormous restaurant the way I see it."[1] I prefer to think of nature as a buffet: animals can pick and choose from among the living things in whatever part of the biosphere they inhabit. Items on this "biospheric buffet" are constantly being swapped in and out with changing environmental conditions. For example, fruits and leaves are replaced by grass roots and tubers as forest gives way to savanna when and where the climate becomes cooler and drier. Different habitats mean different options, choices, and relationships between a species and its environment. In short, a species' choices help define its relationships with other organisms, both eaters and eaten, and its place within the larger community of life that surrounds it.

This book tells a story of teeth, diet, and human origins. My goal is to show that we can use teeth to understand the diets of our ancestors, and, by extension, our place in nature and how we came to be the species we are today. Central to this story are the effects of climate and environmental change on our ancestors, a story only now coming into focus as scientists begin to understand that our success as a species is due, in no small measure, to how our ancestors dealt with an increasingly variable and unpredictable world in the distant past.

When we bring together these new insights from Earth system science and paleoclimate studies, along with new approaches to how teeth work and new discoveries in paleontology, primatology, archaeology, and other fields, we arrive at a more complete view of life in the past and how it changed over time. The story used to be simpler. The spreading savanna coaxed our ancestors down from the trees, and the challenges it brought made them human. It is now becoming clear, however, that environmental conditions actually swung back and forth between wet and dry in the past. This fluctuation winnowed out the pickier eaters among us, leaving only those flexible enough to find something with which to fill their plates from an ever-changing biospheric buffet. We are the most versatile of primate species, able to find something to satiate us no matter where we roam. That explains how we came to take over much of the world. Climate change provided the motive, and evolution offered the opportunity to make us human.

———

I decided to write this book to help me see the big picture in human evolution, and to share what I've learned from being involved in the work for the past three decades. But as I began to think about it, it became clear to me that there's much more to the tale than the science itself; there's also the passion, ingenuity, and determination of those who gave us the knowledge we have. They make the story compelling and bring it to life. And so, in the chapters that follow, we travel around the world, visiting with my colleagues and other scientists along the way. We look over their shoulders as they make their discoveries and chart new paths to understanding the past.

We begin with teeth, how they work and how they are used. If we assume that nature selects the best tools for the job, animals with different diets should have teeth to match. Understanding how teeth work is the first step in figuring out relationships between dental form and function. And working out those relationships is fundamental to using teeth to reconstruct diets of fossil species. But it's not enough. We know from watching living primates in their natural habitats that food choice is about much more than what an individual is capable of eating. The difference between how teeth work and how they are used is key to understanding diet, place in nature, and ultimately evolution. This is our point of departure from business as usual in paleontology, the assumption that animals are fated to specialize on the foods to which their teeth have evolved. Yes, every species is limited by its anatomy in what it can eat, but it's equally important to remember that the dishes on the table vary from place to place, and that they get swapped out from time to time. In other words, food choice is not just about dietary adaptation but also about availability. When the options change, so too does diet and, along with it, the relationship between an organism and its environment.

With lessons learned in the laboratory about how teeth work, and in the forest on how they're used, we move forward into the past. We consider the cast of characters in human evolution, both the fossil species and those who worked to find and make sense of them. The idea that human evolution was somehow triggered by our changing world is not a new one. But as scientists sort out the details of Earth's climate history and reconstruct ancient environments, our old model of human ancestors descending from the trees to meet a spreading savanna falls like a house of cards in a stiff wind. We're taught that the dithering tilt of the Earth and its orbit about the Sun set the pace for climate change, and that shifting continents transform the face of our restless planet. The cradle of humankind didn't simply become cooler and drier; conditions wavered back and forth with increasing intensity over time. Our ancestors met a more and more unpredictable world, and it was their responses that drove our evolution.

What were those responses? With what did our ancestors choose to fill their plates as nature swapped dishes on the biospheric buffet? The sizes and shapes of fossil teeth offer some clues, but those are more about what our ancestors could eat than what they actually ate on a daily basis. We have to turn to what I call *foodprints*, actual traces left by foods eaten during life. Distinctive patterns of tooth wear and the chemistry of dental tissues help us fill the gaps. Only then can we begin to understand the roles of our ancestors in their larger communities of life, and their places in nature.

We can also use our new approach to explore other great transitions in human history. How did a changing world make us human? That's the story of the origin of our biological genus, *Homo*, and the start of the hunting and gathering lifestyle that gave our ancestors the versatility they needed to spread across the planet. And how did we change the rules of the game and begin to stock the buffet ourselves? That's the story of the Neolithic Revolution, the shift from foraging to farming. Monkeys and apes don't work as models anymore. We must look to the few peoples that today still eat wild plants and animals, those whose ancestors never jumped on the food-production bandwagon. Archaeologists have moved countless tons of dirt to document and explain these transitions too. To make a long story short, both are tied to environmental change, and both left marks we can decipher on the teeth and bones of our ancestors.

This takes us up to the present but, in an interesting twist, right back to the past. Paleolithic diets are hugely popular today, and they bring attention to the kind of research I do. Many argue that there is a mismatch between our diets now and those our bodies have evolved to eat. They believe that this explains most of the chronic degenerative disease plaguing our health care systems. In effect, the adaptations for dietary versatility that led to our success have at the same time made us a victim of it. We *can* eat much more than we should. And while I am not a fan of Paleolithic diets for the simple reason that there is no single ancestral diet to which we evolved, there's little doubt that an evolutionary perspective can teach us a lot about our bodies and their welfare.

The book ends where it began, with teeth. Other species don't have crooked, crowded, and impacted teeth riddled with holes. Why do we? It is clear that, while our ancestors ate different foods at different times and in different places, there is a genuine mismatch between our diets today and our teeth. If we consider them in this light, they remind us of our evolution. Our teeth connect us to our ancestors.

CHAPTER I

How Teeth Work

I took my daughters to the Museum of Discovery in Little Rock when they were young. They couldn't have been older than five and seven. In a small display cabinet at the back of the museum, we found two skulls lying side by side: a giraffe and a lion. I asked the girls, "Which one is the meat eater?" They both looked at the teeth and pointed to the lion. The giraffe has broad, flat molars for grinding leaves, and the lion has sharp, bladelike ones for slicing meat. They knew this intuitively. The differences between the teeth of a herbivore and a carnivore are obvious. What about subtler differences, say between animals that eat different parts of plants or those that eat different parts of animals? What about our teeth? What kinds of diets are they designed for?

These are important questions to paleontologists, especially to those of us who work with the fossilized remains of early humans. Our job is to document and explain the course of evolution and to reconstruct life in the past. We often have little other than teeth with which to do it. When you think of fossils, you might envision the massive skeleton of a *Tyrannosaurus rex* or a mammoth dominating a capacious hall at a natural history museum. You might think of the skull of a human ancestor on the cover of a popular science magazine, posed for striking effect and lit to reflect our shadowy distant past. But such fossils are actually few and far between.

There are often hundreds if not thousands of teeth for every skeleton or complete skull we find. That's because teeth are essentially ready-made fossils. The enamel that coats ours, for example, is 97% mineral, with the rest water and trace amounts of organic material. Teeth are stronger than bones, and they are much more likely to survive the ages. Fortunately for paleontologists, they are also excellent tools for understanding life in the past. If we can reconstruct diet from teeth, for example, we can use them as a bridge to the worlds of our ancestors and other long-gone species. And the more about diet we can wring from fossil teeth, the better the resolution with which we can see those worlds.

But to get there, we need to understand how teeth work. Researchers have been grappling with that task for more than a century, and this chapter chronicles the process. How did the countless variety of mammalian tooth types, including ours, evolve from the modest peg-like structures of most fishes and reptiles? How does this relate to changing tooth function from simple biting to complex chewing actions? How do teeth actually break food? How does change in tooth shape with wear affect the process? One question has led to the next. New finds, new techniques, and new theoretical approaches have come and gone, contributions of the scientists who have stretched the limits of our knowledge built upon the foundation laid by those who came before.

EVOLUTION TO AND FROM THE MAMMALIAN MOLAR

We could begin the story with the discovery of the earliest teeth, which date back nearly half a billion years. Most were simple, pointed structures used to capture and immobilize prey, and to scrape, pry, grasp, and nip all manner of living things. These gave their bearers an advantage in the "evolutionary arms race," as Richard Dawkins called it. The better an eater was at getting energy and other nutrients, the more babies and the more evolutionary success it had. Teeth spread quickly through the primordial oceans, and the fish lineages that had them eventually sidelined the groups that did not.

But to understand the nuances of how our teeth work, we need to start much later,[1] with the earliest mammals and their immediate predecessors. We need to start with the transition from teeth used mainly for obtaining food to those used for chewing it. Chewing separates food into pieces small enough to swallow, breaks open protective casings that would otherwise pass through the gut undigested, and fragments morsels to expose more surface for digestive enzymes to work on. This gave early mammals the extra fuel they needed to generate their own heat, which meant they could be active at night and live in colder climates or those with more fluctuating temperatures. They could sustain higher levels of activity and travel speeds to cover larger distances, avoid predators, capture prey, and produce and care for offspring.

The ability to chew opened a whole new world of possibilities for the earliest mammals. And today they come in a dizzying variety of shapes and sizes, from the tiny bumblebee bat, at less than a tenth of an ounce, to the behemoth blue whale, at nearly 200 tons. They are found in an incredible variety of habitats, from Arctic tundra to Antarctic pack ice, high-altitude mountaintop to deep ocean water, and desert to rainforest. We are members of a remarkably successful biological class of animals. That success is due, in no small measure, to the ability to heat the body from within, and the chewing that gave our common ancestor the fuel it needed for the task.

Pay attention the next time you eat something. Your jaw, tongue, and teeth act in concert with sensory feedback. The alignment and movements of opposing teeth are precise to 1000ths of an inch as you generate, direct, and dissipate the forces needed to break food. You position and hold objects in your mouth and keep air and food passages separate to prevent choking. All this is tightly coordinated, with the various parts working together in symphony and synergy.

How could such an amazing system evolve? The transition is detailed in a fossil record of mammal-like reptiles and early mammals that spans more than 100 million years. The changes to the jaw and mobility of its joint; reorganization of the muscles that move that jaw; development of a bony palate to separate air and food passages; and differentiation of teeth into incisors, canines, premolars, and molars that fit together precisely enough for chewing are all there. The transformation culminated with a new and very special kind of molar tooth, the one from which ours and those of other living mammals ultimately evolved. It's with this molar that our story really begins.

The Missing Link between Reptilian and Mammalian Teeth

The basic model for the evolution of mammalian molar teeth was worked out by Edward Drinker Cope, one of the most productive and colorful characters in nineteenth-century paleontology.[2] Cope was born in 1840 just outside of Philadelphia, the son of a wealthy merchant.

His father had expected him to become a gentleman farmer, but Cope was much more interested in natural history. While his formal higher education was limited to a single course in comparative anatomy at the University of Pennsylvania, he racked up countless hours of practical training, cataloging the reptile collection at the Academy of Natural Sciences, with frequent visits to the Smithsonian in Washington, DC, for comparative study. Cope also visited natural history collections around Europe, and called on some of the most important naturalists of the day there. His father, a devout Quaker and pacifist, had sent him to Europe in 1863, in part to keep him from enlisting or being drafted into the American Civil War.

Cope met Othniel Marsh in Germany. Marsh was an American graduate student studying paleontology at the University of Berlin at the time. Their relationship started amicably enough, and both returned after a time to the United States to begin working as paleontologists. Marsh's wealthy uncle, international merchant and financier George Peabody, founded the Museum of Natural History at Yale at Marsh's behest. Marsh became its first director. Cope went to teach at Haverford College as professor of zoology,[3] though he quit his job there after just a couple of years, evidently to focus on research. In fact, Cope moved his family to Haddonfield, near important dinosaur beds in western New Jersey. But trouble started when Marsh came to visit in 1868 and made a deal with quarrymen there to send whatever new fossils were found to Yale. Cope was incensed. To make matters worse, Marsh soon after called Cope out in public for reconstructing the skeleton of an extinct marine reptile incorrectly. To his embarrassment, Cope had put the head at the end of the tail rather than on top of the neck.

These events touched off the "Great Bone Wars," perhaps the greatest rivalry in the history of science. Competition between the two men was fierce as each jockeyed for position in the field and worked to destroy the other's career. A dark cloud hovered over American paleontology in the late nineteenth century, but there was a silver lining. Cope and Marsh were both incredibly productive as each poured his energy and resources into besting the other. Cope alone published more than 1400 scientific works and amassed 13,000 new fossil specimens, despite his untimely death at 57. Most important for us, this all-consuming rivalry led to the discovery of how mammalian teeth evolved.

In 1872 the US Congress authorized a series of expeditions to map parts of the country west of the 100th meridian. Cope joined the survey two years later, ostensibly to chart geology. The Bone Wars were on, and this would be a great opportunity to search for new fossil sites in an area Marsh had not worked. Among Cope's many finds was a series of layers of lime- and clay-rich rock in badlands near the town of Cuba in northwestern New Mexico. He separated the layers into two formations he named the *Puerco* and *Torrejon* and recognized them to be just younger than underlying deposits known to contain dinosaur fossils. These are today combined into the Nacimiento, which formed early in the Paleocene epoch, between about 65 and 58 million years ago,[4] just after the rock fell on the Yucatán to end the reign of the dinosaurs. Cope didn't find any fossils in the formation at the time, but the Nacimiento would later become known as, in the words of his junior colleague Henry Fairfield Osborn, "the most unique and important palaeontological discovery of his life."

Shortly after Cope published his study of the geology from the survey, Othniel Marsh hired a frontiersman named David Baldwin, who had also participated in the mapping expeditions, to collect fossils in the area. Baldwin and his more-or-less faithful burro wandered northwestern New Mexico on and off over the next four years in search of specimens for Marsh.[5] He was untrained and wrote poorly, but was very good at following rock strata and finding fossils. Baldwin found and sent dozens of boxes of weathered bits of bone and teeth to Yale. But Marsh wasn't impressed with the scraps and, after a dispute over payment for his efforts, Baldwin quit. He began working for Cope instead, and continued to do so for the next eight years.

While Marsh hadn't recognized the importance of Baldwin's finds, Cope did, especially those from the Nacimiento Formation. These were the first Paleocene-age land animals found in the Americas and the very best record anywhere of primitive mammals alive just after the dinosaurs. Nearly everything Baldwin found was new to science, and Cope described more than 100 new species of vertebrates from the Puerco series. They were incredible finds—and Marsh had missed them. These bits of bone and teeth turned out to be more important, at least to those of us who care about fossil mammals, than all the big, beautiful dinosaurs the Bone Wars ever produced. Baldwin had found a link between the

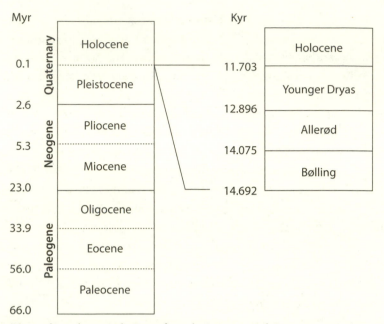

1.1. The geological time scale. Dates from the International Commission on Stratigraphy, "International Chronostratigraphic Chart v2015/01," http://www.stratigraphy .org/ICSchart/ChronostratChart2015-01.pdf. "Myr" refers to millions of years ago, and "Kyr" refers to thousands of years ago.

humble cone-shaped teeth of reptiles and the myriad forms of molars adorning the mouths of living mammals. Cope had been trying to understand how complex teeth evolved from simple cones for more than a dozen years. Baldwin's new fossils gave him the key to deciphering the evolution and diversity of mammalian molar types.

Cope described the teeth at a meeting of the American Philosophical Society in 1883. Most of the upper molars had triangular-shaped biting surfaces with three main cusps, or tubercles: two on the cheek side and one on the tongue side. He called these teeth *tritubercular*. A row of upper molars is basically several of these triangles lined up end to end so their bases are continuous and their tips all face the same direction. The cusp at the tip of each tooth is connected by a sharp ridge to the one in front and the one behind, forming a "V" that points in toward the tongue. When you line up the teeth, the crests form a continuous zigzag-shaped blade running the length of the row, kind of like a jagged cookie cutter (e.g., VVV).

The uppers are paired with lowers that have matching triangles facing the opposite way (e.g, ΛΛΛ), with two cusps on the tongue side and one on the cheek side. The lower triangles fit between the uppers, so that opposing blades slide or shear past one another. Baldwin's Paleocene teeth were amazing slicing and dicing machines that could give the average late-night infomercial food processor a run for its money. But wait, there's more! The lower teeth also had a low shelf that formed a basin to oppose the inner cusp of the uppers. Food could be crushed and sheared at the same time. The tritubercular tooth was a dazzling feat of engineering, and a link that indeed allowed Cope to connect the chain between the primitive cones of our reptilian ancestors and the specialized teeth of mammals that followed.

Cope speculated that early ancestors of the mammals started out with cone-like teeth, and that small cusps were added in front and back. Over time, nature rotated the new cusps out of line from the original cone, outward for the uppers and inward for the lowers, to form the reversed triangles of opposing rows. A shelf set on the back end of the lowers completed the effect. Cope argued that it was easy to build the teeth of today's mammals from this basic form. Straighten the crests and take out the shelf, and you've got the bladed teeth of cats and dogs. Add a fourth cusp to square off the triangles and raise the shelf, and you've got our molars. A few additional tweaks get you to a horse tooth or a cow tooth. Step by step, then, Cope traced the path of evolution from reptilian tooth cones to the ancestral mammalian molar to the myriad forms we have today. His was a brilliant theory, based on intuition, reason, and fossil evidence.

More than half a century later, paleontologist William King Gregory would write, "One can only wonder at Cope's amazing insight into the problem of the evolution of the dentition. All that has been done since is practically only an amplification and verification of this prophetic passage."[6] Imagine trying to assemble a puzzle with no picture to guide you, most of the pieces missing, and no idea of how those you have fit together. This was harder. And Cope got it basically right. His model was soon confirmed by his junior colleague, Henry Fairfield Osborn, then professor of comparative anatomy at Princeton. Osborn studied even older, more primitive teeth of mammals from the Mesozoic era, the age of the dinosaurs. This also allowed him to fill in some of the details Cope's model was missing. Today we can quibble about a few of the

A

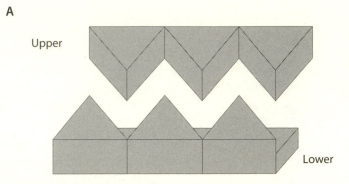

B

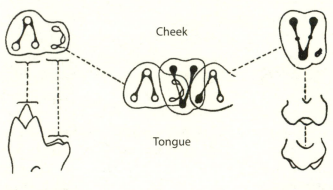

1.2. The tritubercular, later referred to as the tribosphenic, molar. A. stylized reverse-triangle configuration of upper and lower molar teeth; B. illustration of upper and lower molars, modified from Henry Fairfield Osborn, *Evolution of the Mammalian Molar Teeth to and from the Tribosphenic Form* (New York: Macmillan, 1907).

specifics of the Cope-Osborn model, particularly some of those introduced by Osborn, but it remains the foundation upon which our understanding of the origin and evolution of mammalian molar teeth is built.

THE ROLE OF TEETH IN CHEWING

Neither Cope nor Osborn spent much time worrying about the details of how fossil teeth worked in life, or what sorts of foods they were used on. The early days of paleontology were about digging, describing, and naming fossils. Then came arranging species into groups, figuring out relationships between them, and tracing the evolution of specific anatomical forms, like the tritubercular molar, through time. But identifying that molar type, and working out how it developed from a simple cone and then evolved from there, is integral to the story. Without those details, researchers could not have begun to understand the role of teeth in the evolution of mammalian chewing. Nor could they have figured out why teeth work as they do today.

Fossils as Animals Alive in the Past

George Gaylord Simpson knew this intuitively, even back when he was a graduate student at Yale in the 1920s. His adviser, Richard Swann Lull, was the new director of the Peabody Museum there and a former student of Osborn's. Like his academic grandfather before him, Simpson chose to work on Mesozoic mammals. There was a great collection assembled by the museum's first director, none other than Othniel Marsh, during the Great Bone Wars. Many of the teeth and jaws had never been properly cleaned, prepared, or described, and Simpson wanted the job. Lull was initially not happy with the idea, but he acquiesced after several months, and Simpson wrote a series of papers on the specimens, many while he was still in graduate school. It was an interesting twist of fate, Simpson building on Cope's legacy by working on Marsh's fossils!

The fourth paper in the series was especially important. Simpson introduced a whole new way of looking at fossils. In his own words, it was an "attempt to consider a very ancient and long extinct group of mammals not as bits of broken bone but as flesh and blood beings."[7] The approach seems obvious to us today. Fossils are the bones and teeth of animals that were alive in the past, after all. How else would you understand them other than by applying the principles of biology? But most paleontologists at the time weren't thinking of fossils as once-living organisms. In fact, it had been only a few years since the term *paleobiology* was coined, evidently by Lull's immediate predecessor at the Peabody Museum, Yale professor Charles Schuchert.[8] Perhaps he had influenced Simpson.

In any case, Simpson used tooth shape and direction of scratches on wear surfaces to figure out how uppers and lowers fit together and how they moved during chewing. He compared his study animals, a group called the multituberculates, to living species. There are no animals alive today quite like the multituberculates—their molar teeth have two or three rows of up to eight cusps each, arranged front to back. But they do have bladelike, serrated (like a steak knife) premolars very similar to those of the Australian rat kangaroos. So Simpson figured that, like rat kangaroos, the multituberculates must have used their back teeth to grind tough vegetation in life. The pattern of wear scratches on their molars showed that the rows of cusps on the lowers slid front to back between those on the uppers during chewing. These teeth would have made an excellent milling machine for fibrous plant parts. They must have worked well too, because the lineage lasted for more than 100 million years. And the multituberculates dominated their landscapes for much of that time.

When Simpson finished his PhD research, he spent a year studying Mesozoic mammals at the British Museum in London and at other natural history collections in Europe. He returned to the United States in 1927 to take a job at the American Museum of Natural History in New York City. His academic grandfather, Henry Fairfield Osborn, was president of the museum at the time. Simpson developed his ideas on tooth form and function over the next few years, and laid them out in detail in his 1933 paper, "Paleobiology of Jurassic Mammals." Simpson was later to become the most influential paleontologist of the twentieth century but, to me, this was the most important of his nearly 400 academic works. He wrote, "change of dental form, in itself, is not the important point in considering mammalian evolution. The mammals were living creatures, and the teeth are not inorganic objects but a means for obtaining and utilizing nourishment." He continued, "this paleobiological approach to the subject infuses it with life and gives real meaning to the study of teeth."[9]

In that 1933 paper, Simpson laid the foundation for our understanding of how teeth work. He first reasoned that chewing is a product of both jaw movement and tooth shape, and then differentiated vertical from horizontal motions, and teeth with opposing crests from those with cusps and basins. He called vertical movement of crests sliding past one another *shearing*, horizontal movements of cusps sliding across basins *grinding*, and vertical movement of cusps into basins *opposition*. He then related each of these *occlusion types*, as he called them, to a different diet: shearing for meat, grinding for plants, and opposition for a combination of the two. Truth be told, Simpson's model was a bit more

complicated than this, but his basic take-home message was simple—
teeth are guides for chewing. So we can use the shapes of fossil teeth to
work out how past mammals chewed and what they ate. Researchers
ever since have focused on the details.

Simpson also coined the term *tribosphenic* to replace *tritubercular*.
The new name underscored the change in approach from simply documenting tooth shape, three tubercles or cusps, to thinking about function. *Tribein* is Greek for "to rub, grind, or pound," and *sphen* means
"wedge." "Tribo-" was for the action that Simpson described as like a
mortar and pestle between the inner cusp of Cope's upper triangle and
the shelf on back of the lower. "Sphenic" was for the way each lower
triangle fit as a wedge between the uppers for shearing between opposing blades. It was a great multipurpose tool—shearing, crushing, and
grinding in one tooth. Like Cope and Osborn before him, Simpson
recognized the tribosphenic molar as "the common denominator, the
point of departure" from which the teeth of all living mammals arose.

Simpson and his contemporaries spent years matching wear facets on
opposing fossil teeth to work out exactly how they fit together. They considered the directions of scratches on those facets, mobility of the jaw
joint, and where chewing-muscle attachment sites were on the skull to
reconstruct tooth movements. Over the next couple of decades, researchers looked at more and more fossil species. Some went back to the reptilian ancestors of the earliest mammals. Others pushed forward, tracing
individual lines toward today's groups. The evolutionary history of mammalian teeth began to take shape, and paleontologists speculated on how
chewing and diet changed over time within and between lineages. But
not everyone came up with the same answers. Researchers had taken the
approach about as far as they could at the time. What was needed was a
better understanding of how teeth fit together and move relative to one
another during chewing. No one had actually documented how living
animals break down food during the process. Of course, that is easier said
than done with cheek, lips, and tongue in the way.

Bridging the Gap between Fossils and Living Mammals

Bridging the gap between fossils and living animals would take a part-
nership between a paleontologist who recognized the need to do so and

an experimental biologist with the skill set to pull it off. Fuzz Crompton had studied mammal-like reptiles from the early Mesozoic for his PhD at Cambridge in the early 1950s. He returned to his native South Africa in 1954 and spent the next decade working on the origin and evolution of mammalian jaws and teeth at museums in Bloemfontein and later Cape Town. He accepted the directorship of the Yale Peabody Museum in 1964 and stopped in London on his way to New Haven. That stopover would be a defining moment, both for his research career and for our current understanding of how teeth work.

Crompton was visiting a friend who had just heard a talk by Karen Hiiemae, then a graduate student at St. Thomas' Hospital Medical School. Hiiemae was pioneering the use of X-ray motion pictures to study jaw movements in rats. What about teeth? Could this new technique be used to watch them in action too? If so, maybe he could see exactly how opposing teeth come together during chewing. This could be a great way to build on Simpson's work, to understand the effects of tooth shape, and how wear facets form. Crompton arranged to meet Hiiemae, and the two began a collaboration that would span four decades. They set up an early cine X-ray camera in the basement of the Peabody Museum in 1967 and began in earnest to watch animals chew.

They turned first to the opossum, a primitive-looking mammal, vaguely ratlike and about the size of a house cat. It's an American marsupial whose range spans much of the western hemisphere and which eats just about anything it can get in its mouth, from berries and nuts to insects and small vertebrates. Cope and Simpson had both recognized well before Crompton and Hiiemae came along that opossums had basic tribosphenic molars very much like those they had described for the ancestor of today's mammals. This animal was in a sense, then, a living fossil, the perfect starting point for a study of mammalian tooth function. If Crompton and Hiiemae could watch opossums chew, and relate tooth shape to jaw movements, they could understand how the molars of early mammals worked. From there, it should be a short step to figuring out how more specialized teeth relate to jaw movements, and, ultimately, to understanding the evolution of different types of mammalian molars.

Crompton and Hiiemae discovered that the opossum uses its teeth in two ways. It first tenderizes or pulps food by puncturing and crushing with its pointed cusp tips, and then slices it between crests formed on the slopes of those cusps. The lower tooth crests take the shape of a V in side view, and the upper ones form a Λ, so that when opposing teeth slide past one another, their blades form the walls of a shrinking diamond and the food between them is trapped and sliced without spreading. It's kind of like a double-bladed cigar cutter. The multiple cutting planes and sharp, piercing cusps make a great general-purpose tool for the varied diet of the opossum.

The cine X-ray studies also showed that, as Cope and Osborn and Simpson had speculated decades before, the tribosphenic molar makes for a great jumping-off point for more specialized tooth types. Dogs and cats, for example, have consolidated the many small shearing surfaces in their ancestor's mouth into a few large ones on which to concentrate their effort to slice tough animal tissues. Crompton and Hiiemae pointed to the lynx, with its long, sharp blades running front to back on the last upper premolar and the first lower molar. These *carnassials*, as they're called, allow carnivores to cut large chunks of flesh in a single stroke. There is little side-to-side movement during chewing, because their blades run the length of the tooth rather than across it, as we saw in the tribosphenic molar. The system works more like a guillotine—up and down.

The molars of herbivores also evolved from the tribosphenic form, but in another direction. While meat eaters need only slice food so that it's small enough to swallow, tough, fibrous plant parts must be broken down more before they reach the gut for digestion. Carnassials simply don't cut it. As Crompton and Hiiemae suggested in one paper, watch a dog try to eat a grass blade and you'll understand. Herbivores typically have more square-shaped teeth with a broad and level biting plane cut by rows of low crests connecting small cusps. Think of a washboard or a file, with closely packed, raised ridges forming a roughened surface. Vegetation is ground or milled between opposing teeth when the lowers slide along the uppers in the direction opposite the length of the crests. For cows and sheep, those crests run front to back, so they chew side to side. For many rodents, the crests run sideways across the crown, so chewing is back to front. It's the same fundamental idea, just imagined in a different way.

Primates Enter the Picture

Opossums, lynxes, cows, sheep, and rodents are great, but this book is about us. If we are interested ultimately in how *our* teeth evolved, we need to understand how primates chew too, and how the subtle differences in tooth shape between species relate to diet. Our nearest living relatives don't have primitive tribosphenic molars, bladed carnassials, or complicated milling teeth. Monkeys and apes have simple, rectangular tooth crowns with four or five cusps each, in some cases blunt and rounded, in others sharp and pointed. Crests run up and over the cusps, front to back, on both the tongue and cheek sides of the crown. And basins can form between them, enclosed by the cusps.

The task of working out those details fell to another Yale graduate student, Rich Kay, in the early 1970s. Kay didn't start out interested in building on Simpson's legacy, or even in working with Fuzz Crompton. Crompton had actually just left Yale before Kay arrived, to head up the other Peabody Museum, the one at Harvard. Kay went to Yale with the idea of working on primate fossils in the quarries of Egypt's Western Desert. But, to paraphrase Robert Burns, the best laid plans of mice and men often go awry. Shortly after Kay arrived in New Haven, some members of the Yale field crew inadvertently stumbled into an Egyptian military encampment. It was in the wake of the Arab-Israeli Six-Day War, and the government responded by closing the quarries to scientific research.[10] Kay needed a new project.

As luck would have it, that was just the time that Crompton returned to New Haven for a visit to give a series of lectures on mammalian chewing. Kay was hooked. He scored an invitation to visit the new Crompton Laboratory at Harvard and soon began working there with Karen Hiiemae. The two spent much of the next couple of years watching monkeys crush, shear, and grind different sorts of food to document the process. As with opossums, primates sliced food between the leading edges of shearing crests connecting the front and back cusps as opposing teeth slid past one another. Food was crushed as cusps of the lower teeth pressed into the basins of the uppers and vice versa. It was ground by a combination of compression and drag between the teeth.

Kay was interested in more than just how primates chewed, though. He wanted a link between tooth shape and diet. This could be a great

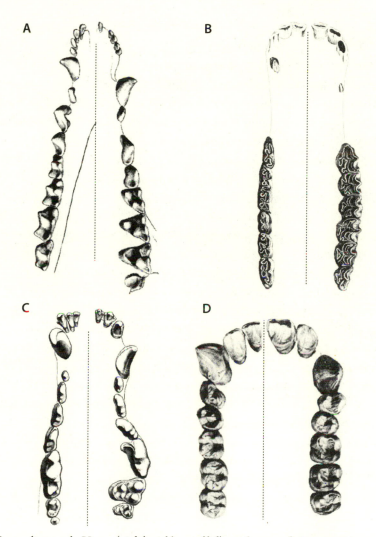

1.3. Mammalian teeth. Upper (*right*) and lower (*left*) tooth rows of: A. an omnivore (opossum); B. a herbivore (horse); C. a carnivore (wolf); and D. a frugivore (chimpanzee). Images modified from Christoph Giebel, *Odontographie: Vergleischende Darstellung des Zahnsystemes der Lebenden und Fossilen Wirbelthiere* (Leipzig: Verlag von Ambrosius Abel, 1855) (A. and C.), and Richard Owen, *Odontography* (London: Hippolyte Bailliere, 1840) (B. and D.).

bridge between his work at Harvard and his interest in primate fossils. Think of the implications for paleobiology. He would have to take measurements, like the lengths of shearing crests and area available for crushing and grinding. Longer crests, more shearing. Larger basins, more crushing or grinding. He would need the teeth of primates with known diets, and lots of them. That would take him into the museum collections at Yale, Harvard, and the American Museum of Natural History.

The public areas of natural history museums are dominated by enormous open halls with stuffed elephants on display, *Tyrannosaurus rex* skeletons, and 10-ton meteorites. But there's a whole other world behind the door signs that read, "Staff Only." Curators often have small, cramped offices filled with papers and books piled high or strewn about haphazardly. Their job is to maintain huge collections of specimens, which can leave precious little energy to organize much else. Most visitors get to see less than five % of the collections. The vast majority of primate skulls, for example, are usually kept in small boxes clustered on shelves in cabinets. In most museums they are piled from floor to ceiling, lining large rooms, small rooms, and the hallways between them—anywhere they can be fit. The stream of cabinets is broken only by well-worn tables where visiting researchers sit and do their work. Faded white walls behind them are a stark contrast to the colorful display halls on the other side.

Kay spent countless hours at those tables, staring down the barrel of a microscope at tiny tooth parts. He measured shearing crests, basins, and other features on thousands of primate molar teeth. Then he began to plot the data. Tooth length on the horizontal axis, crest length on the vertical one. The separation was clear. Species that eat leaves or insects have longer shearing crests relative to the length of a tooth than those that eat fruit, just as he had expected. It makes sense. You'd choose a blade to break a leaf into pieces, not a hammer. It would be a different story, of course, if you were trying to break something hard and brittle, or soft and pulpy. But more on that later.

What Kay had done was to develop an objective, quantitative way to reckon the diets of long-dead primates from the shapes of their teeth. Measurements for a fossil tooth could be plotted on the same graph as those for living species with known diets. A fossil specimen with tooth and crest lengths that fell in with fruit eaters was probably one too. His method, now called *shearing quotient analysis* in the textbooks, quickly became the standard for measuring teeth in paleontology laboratories the world over, and it's still used widely today.

Kay's approach is what we call *bottom up*, or inductive reasoning. Most historical science is not about lab-coat-clad researchers mixing chemicals in beakers or scribbling notes on a clipboard as a mouse navigates a maze. You can't run experiments on things that happened in the distant past. The best we can hope for is to uncover basic principles and relationships that work today, and assume that these haven't changed over time. Historical science is about developing and testing hypotheses, like "monkeys with long shearing crests eat leaves." We can't know for sure if this has always been true, but we can look at primates alive today and try to prove it wrong. If only leaf eaters have long crests,[11] and none have short ones, we're probably on to something—especially since the mechanics of it make sense. So we test hypotheses about form-function relationships on living species, and build the confidence to infer function from form by failing to reject them.

THE ROLE OF TEETH IN FRACTURING FOOD

So far we've focused on chewing in terms of the way teeth fit together, and the way they fit together in terms of their shape. This implies that teeth function principally as guides to direct chewing movements. Crests form to slide past one another in shearing, while cusps are there to press into basins for crushing. From the jaw's perspective, then, teeth are basically passive players in the chewing game.[12]

On the other hand, chewing is most fundamentally about breaking down food, not moving the jaw. If we look at the process from the food's perspective, the teeth are not passive at all. They're the active players. While this may seem like a matter of semantics, it's most certainly not. Starting with food and working back to teeth, rather than the other way around, is a fundamentally different way of thinking about chewing. When dental researcher Peter Lucas introduced the approach, it shook the very foundation of our discipline. Lucas started with first principles and took a more top-down, or deductive, approach to understanding the process. He used the physical laws of food fracture to develop hypotheses; in this case, "idealized" teeth for breaking down foods with specific properties. These could be tested by comparing the models with real-world teeth to see how closely they matched.

Lucas got his start studying physical anthropology as a college student in London in the 1970s. He was interested in teeth and human evolution, but felt like a frustrated physical scientist, more comfortable in the company of physicists and engineers than among anthropologists. Surely, physics could be used to work out how teeth break food, and that knowledge could be applied to reconstruct diets of fossil species from the shapes of their teeth. He led the charge with his dissertation, a first effort to apply principles of mechanical engineering to how primate teeth break food.[13] And he spent the next three years as a postdoc in the laboratory at Guy's Hospital in London, working out details.

Lucas took his first teaching job in 1983 at the National University in Singapore. The next step should have been to match his newfound knowledge about how foods fracture to his interest in fossil teeth. But there were no fossils to study in Singapore. He had to find something else to do. He soon met Richard Corlett, an ecologist who had arrived on the island the year before. Corlett worked on plants in the Bukit Timah Nature Reserve, a small vestige of rainforest just a few minutes' drive from campus. Lucas and Corlett hit it off, and together began studying the reserve's monkeys to document diet and tooth use during feeding. Everything fell into place. Lucas was in the right place at the right time with the right mind-set and expertise. He would later collect food samples, bring them back to the lab, and figure out how they broke. The studies of feeding behavior and food fracture together were revolutionary and changed the way we think about how teeth work.

Back when Peter Lucas started in the field, researchers understood that the way a food breaks depends on its physical properties. But without a firm foundation in fracture mechanics, it was difficult for them to make much headway on exactly how this worked. Small, tough objects; coarse or fibrous vegetation; hard seeds and nuts—paleontologists were bandying about descriptions of food with a reckless abandon that would make even the most stoic of materials scientists cringe. Terms like *hard, soft, tough,* and *brittle* have precise definitions, and engineers tend to get very upset if they are used wrongly. More important, use of those terms implied an understanding of how specific foods break that simply did not exist. Also, with researchers using them to mean different things, progress did not come easily. Lucas made many of us realize this and, in the process, begin to look at chewing from a food's-eye view.

How Foods Break

How does food break? It's a simple enough question, but one to which most paleontologists haven't given much thought. Unless we can answer it, though, we can't really understand how teeth work. In the simplest sense, breaking food is about starting and spreading cracks. The details are the stuff of fracture mechanics. Let us assume for a moment that most living things would prefer not to have their parts eaten if given a choice. There are exceptions, like fleshy fruits, but plants and animals put a lot of effort into protecting themselves from being eaten. Some run away or fight back; others have evolved toxins to discourage predators.

They also make themselves difficult to break so they cannot be swallowed, or at least not digested well enough to make them worth the trouble of swallowing. This is where fracture mechanics comes in. Plants and animals harden their tissues to the point that a crack won't start, or toughen them enough that it won't spread once started. This is often a compromise because hard foods tend not to be tough and tough ones tend not to be hard. Consider a walnut. Its shell is hard, and it takes concentrated force to start a crack. But it is also brittle. Once that crack starts, it spreads easily. Raw meat, on the other hand, is tough. While it takes little to pierce the surface, the real work comes with forcing a crack through to the other side.

There is much more to food fracture than this, though. How much does it bend before it breaks? How much does it stretch? Will it bounce back to its original shape after you stretch it? Solids have so many properties. They're not always straightforward either. Hardness, for example, is really a combination of several different things. The term is actually so imprecise that its mere mention can cause eye rolls at materials science conferences. That doesn't mean the term is useless to us though, so long as we can agree on its meaning.

How Teeth Break Food

The whole process of chewing seems counterintuitive when you first think about it. Food is broken by pressure between opposing teeth. But fracture involves starting and spreading a crack, which requires that we pull bits of food apart, not press them together. In other words, teeth need to figure out how to use compression to create tension. A couple of examples illustrate how this can be done. We can break a walnut by crushing it with a nutcracker. Squeezing around the middle causes the shell to flex and produce tension at the ends. Try it and watch where it cracks. We can split a log by pounding a wedge into it. The wedge is pressed into the wood, but it creates tension at the point of the advancing crack. These and other examples teach us how foods fracture. We can use the knowledge gained to design an idealized tooth for foods with specific fracture properties.

Let's begin with a hard, brittle food, something crunchy, like a nut or a carrot. It takes work to start a crack, but little to spread it once started. The more concentrated the force, the better. A sharp-tipped nail requires less force to penetrate wood than does a blunt one. A hemisphere or dome makes a good model. It minimizes the point of contact with a hard food to concentrate forces coming from the jaw; but at the same time it reinforces the tooth so it doesn't break in the process. Many tooth cusps look like this. For the opposing surface, a platform is good, especially a concave one to keep food in place. A basin or space between staggered opposing cusps does the job quite well. Think of a mortar and pestle.

Now consider a soft, tough food, something chewy, like rare steak or taffy. The challenge isn't starting a crack, but spreading it. In this case, a sharp, wedge-shaped blade does the trick. It should be thin to minimize the energy needed to push the crack through. We don't need to worry as much about breaking it because tough food tends to spread across a surface with compression, lessening the risk of damage. The opposing surface should ideally also be a blade, but the two should be slightly offset, like a set of shears, so they slide past one another rather than colliding and risking damage. Nature makes blades by forming sharp ridges or crests that connect adjacent cusps.

So an ideal tooth for crushing hard foods would have big, bulbous cusps, and one designed for shearing or slicing tough foods would have long, thin, wedge-like blades. If you combine the two, you can get something akin to Simpson's tribosphenic molar, which combines elements of each and makes for a great multipurpose tool to fracture foods with a broad variety of mechanical properties. When we reexamine primate teeth in this light, crests and cusps aren't so much about guiding chewing movements; they're more about food fracture properties. Long, sharp shearing crests of folivorous primates can serve as wedges to fracture and fragment tough leaf blades that resist the spread of cracks. In contrast, the rounded cusps and thickened enamel caps of hard-object-feeding primates are perfect for concentrating and transmitting forces needed to start cracks without the teeth themselves breaking.

Yes, teeth are tools for fracture. But that doesn't mean that they can't also be guides for chewing. One doesn't exclude the other. Scissor blades cut, but they also direct motion. Likewise, leaf-eating primates have opposing crests that virtually interlock when the teeth are brought together. They can't grind because there just isn't enough leeway for the side-to-side movement. So there is little doubt that teeth affect, if not direct, chewing movements when opposing surfaces come together. It's also clear, though, that at a finer level details are about tooth-food interactions—wedges for tough foods, domed surfaces for brittle ones. Think of a bread knife or wood saw. The blade itself allows for the back-and-forth motion, but the cutting happens at the serrations or teeth. The same goes for a file or a cheese grater. The large, flat surface determines the direction of movement, but the work is done at the scale of the ridges or holes. The renowned French naturalist Georges Cuvier realized more than two centuries ago in his *Essay on the Theory of the Earth*[14] that tooth form reflects both levels. He wrote of cows, sheep, and their kin having flat teeth compared with carnivores to allow horizontal motions, but also alternating bands of enamel and dentin for a roughened surface to grind tough vegetation.

THE ROLE OF WEAR: NATURE AS MASTER SCULPTOR

There is a third dimension to the tooth form-function relationship that we haven't considered yet: time. Whether we think of teeth as guides for chewing, implements for fracturing, or both, we assume that nature selects the best tools available to break whatever it is that a species has evolved to eat. But there's a wrench in the works. Teeth wear, and their shapes change over time as a result. This has to affect the form-function relationship. Rich Kay limited his shearing crest studies to unworn teeth for this very reason.

One species may start out with sharper teeth than another, but what happens as they wear and get flatter throughout life? If there is an "ideal" tooth form for a given type of food, how does nature deal with the fact that teeth change over time? Teeth begin to wear as soon as they come into use, and natural selection doesn't just stop at that point. Teeth have to last a whole lifetime, and nature has to take that into account during the design process. We can limit our studies to unworn teeth, but these are few and far between in the fossil record. Besides, unless we consider tooth wear in the form-function story, we get only the first chapter. We miss the best part.

I've argued that nature uses wear as a sculptor would use a chisel to refine and enhance teeth and keep them in good working order as long as possible. Our genes should make teeth predestined to wear down in a specific way, adapted to a specific abrasive environment and diet. In fact, some species actually need their teeth to be worn for them to work properly. Many rodents begin to grind their teeth in the uterus because their molars need to be worn and ready to go shortly after birth. They depend on wear to provide the sharpened, jagged surfaces used to grind tough vegetation.

Nature is an inspired engineer and has come up with a brilliant way to strengthen a tooth and guide wear at the same time. Our crowns have two hardened layers: enamel and dentin. Both are strong and resist wear, but the enamel layer is harder, about 97% mineral, whereas dentin is tougher, 70% mineral but also 20% collagen fibers (think of your tendons, ligaments, and skin). Together, the layers give teeth both the strength and the flexibility they need to break food without themselves being broken in the process. They do it over and over again, up to millions of times during the course of a lifetime.[15] It's a superb design, half a billion years in the making.

But this tissue layout determines more than tooth strength; it also controls how a tooth wears. Imagine a bed with four bedposts, one at each corner. Now, drape an oversized, thick canopy over the posts and allow it to sink downward in the middle. That's a tooth. The bed and posts are the dentin, and the canopy is enamel. As a tooth wears flat, dentin horns (the posts) are exposed and scooped out to form a pit, because they are softer than the surrounding enamel. The rim of the pit forms a sharp edge that makes an excellent cutting surface. By changing the height of the peaks, the depth of the valleys, and the thickness of the overlying enamel cap, nature can predetermine how a tooth will wear and the form it will take during the process.

Understanding how tooth shape changes with wear is an important step toward understanding the form-function relationship. But documenting the process isn't easy. It takes more than measuring crest lengths. I started working on the problem in the mid-1990s when I took a position at the University of Arkansas. The university was hiring a new biological anthropologist for its growing program, and I needed a job. Applicants were asked to include a statement on how they might use the university's recently founded Center for Advanced Spatial Technologies (CAST). Research at the center was focused on *GIS*. I had to look it up. It stands for geographic information systems, a hot new approach at the time for capturing, analyzing, and managing information organized by location in space. If you want to know the best way to lay out roads for traffic control given directions of city growth, emergency management infrastructure, and layout of a business district, build a GIS. If you want to know the best place to plant crops on a farm given water drainage, soil types, and topography, again, build a GIS.

The more I read about it, the more I became convinced that it could be used to study teeth in a way they had never been studied before. When I applied for the job at the University of Arkansas, I was a research associate at Duke, working on fossil primate shearing crests in Rich Kay's lab. So tooth shape was very much on my mind. I began to think of teeth as landscapes—cusps as mountains, fissures as valleys. Why couldn't tools developed for GIS be used on teeth? Measurements of surface steepness, change in slope, and drainage patterns could help us better understand how they work. I would call it dental topographic analysis.

Most studies of teeth in those days involved comparing distances between fixed points, landmarks on the surface. We chose specific points and measurements because they made sense given our understanding of how teeth work. Rich Kay's shearing quotient was a case in point. But there were limitations. There's more to tooth shape than distances

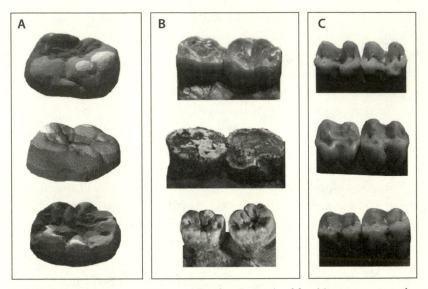

1.4. Dental topography. A. GIS models of molar teeth of fossil hominins *Australo-pithecus afarensis, Paranthropus boisei,* and *Homo erectus*; B. photographs of molars of those same species; C. photographs of molar teeth of a leaf monkey (*above*), a frugiv-orous macaque (*middle*), and a hard-object feeding mangabey (*below*).

between a few pairs of points on a surface. Were we losing important details by limiting ourselves to those measurements?

There was a bigger problem, though. Landmarks on tooth surfaces aren't really fixed points. How can you measure between cusp tips when they change or disappear with wear? Even if landmarks weren't moving targets, different species may not even have all the same ones to measure. Apes have five cusps on their lower molars, but the monkeys that live alongside them have only four. A GIS model could solve the landmark problem by considering whole surfaces. We wouldn't have to decide in advance what parts of the tooth to measure, or worry whether we got the full picture. We could let the teeth themselves teach us how and where they vary with diet. Also, we could compare worn or unworn teeth, and those with different features. Questions I could tackle began to flood my mind. It was exciting.

I started with a small 3D digitizing tablet. Push a tooth into a hunk of clay on the tablet, touch its surface with the stylus tip, and click. Move the pen along the surface, clicking as you go, and you could generate a cloud of up to hundreds of points, depending on your patience, keenness of eye, and steadiness of hand. It was slow going, and the resolution wasn't very good, but it was inexpensive, and the points could be imported into a GIS model.

The next step was to recruit a graduate student who knew something about GIS to help me navigate the labyrinth. Geographic information systems were beginning to take off, but user-friendly software and personal computers powerful enough to make them were still years away. We had to work with cumbersome command-line–based software originally written by the US Army Corps of Engineers for land management and environmental planning. You had to be an expert to navigate it, let alone the computer server and UNIX operating system it ran on. It was pretty intimidating for the uninitiated. For that matter, so was CAST. It occupied several rooms in the basement and ground floor of the geology building; the main lab looked like mission control from an old science-fiction movie, filled floor to ceiling with computer towers, blinking lights, and bulky monitors.

The director at CAST introduced me to a bright young GIS expert, Malcolm Williamson, and we started working together. It is common to suffer frustrations and setbacks when you set out to use technology intended for one thing to do something else. It can also be difficult to get hardware and software created for very different purposes to play nicely with each other. So while the procedures for dental topographic analysis seem pretty straightforward today, the path to developing them was anything but. It was as if every time we worked out a problem, another seemingly intractable one appeared. The digitizing tablet didn't have the resolution needed to document the fine details. So we got a laser scanner that could record a million points per square inch. But the beam went right through the surface; tooth enamel and epoxy casts, even pigmented ones, are translucent. It took weeks of experimentation to find the best way to coat specimens—too thin, too thick, too dark, too light, not sticking to the surface, sticking too well. And it took months to work out all the other kinks and the best measurements to take.

Eventually, though, I could begin focusing on the worn-tooth conundrum. I needed teeth, and hundreds of them. Some unworn, some lightly worn, some heavily worn. I needed the kind of collection kept only in a large natural history museum. But my laser scanner was too bulky and delicate to lug around, and you can't just borrow hundreds of primate skulls. I needed to make replicas, and I needed help. So I asked the best in the business, Mark Teaford, then at Johns Hopkins University. Teaford and I have worked together on primate collections at more than a dozen museums in the United States, Europe, Africa, and South America over the years.

This time it was Cleveland. The Cleveland Museum of Natural History has a wonderful collection of chimpanzee and gorilla skulls, most gathered by medical missionaries working in southern and eastern Cameroon early in the twentieth century. We lined the jaws up on the table, half a dozen at a time, and began to slather teeth with vinyl-silicone impression material. It's the same goo that dentists use to make molds for crown work, and preserves surprisingly tiny details of the original surface. If you can record a symphony in the subtle undulations of the groove on a vinyl disc, why not fine-scale features of a tooth surface in vinyl impression material? The material starts out the consistency of tree sap, but once it hardens it peels right off. The whole process takes just a few minutes, and high-resolution replicas of the teeth can be made by pouring epoxy into the molds back home in the laboratory. Surface features on the cast look just like those on the original, even when they're enlarged hundreds of times.

I enlisted the help of another graduate student, Francis M'Kirera, and we scanned dozens of chimpanzee and gorilla tooth replicas. We expected the leaf-eating gorillas to have steeper, more jagged biting surfaces than the fruit-eating chimpanzees. They did. We expected worn teeth to be flatter than less worn or unworn ones. They were. But things got interesting when we started looking at how differences between the species held up in differently worn teeth.

Chimpanzee and gorilla teeth differed from one another about the same amount in worn teeth as they did in unworn ones. In other words, they maintained their differences through the wear sequence. This was exciting. It meant we could compare worn or unworn teeth between species to get at diet, so long as we compared specimens at the same stages of wear. Even more exciting, the jaggedness of the surfaces hardly differed between unworn and worn specimens. Only the flattest, most worn ones lost their edge. Here was an attribute that seemed not to change with wear. You might think a worn surface loses its jaggedness as it flattens. But once you start breaking through to the dentin, the sharp edge of the enamel compensates. Could this be nature's way of maintaining functionality despite wear?

The Cleveland apes gave us important clues to how nature sculpts teeth with wear, but there's only so much you can do with skulls from a museum. Each individual is represented by a single moment in time, the last of its life. We can string together a series of specimens ranging from unworn to heavily worn to study the wear sequence of a species, but this requires we assume all individuals wear their teeth in the same way. On the one hand, that is all you can do with fossils; each one is effectively a single moment in the life of an individual. On the other hand, we'd have a lot more confidence moving forward if we could prove that different individuals in a species wear their teeth in consistent ways. Otherwise, we can't really argue for species-specific shape changes with wear. I went back to Mark Teaford, but also to his colleague, Ken Glander, from Duke.

Teaford had spent much of the 1980s developing a method to make high-resolution dental replicas of live laboratory monkeys at Case Western Reserve University for his studies of tooth wear. Human patients are easy. The dentist cleans your teeth, dries them, and covers them with a plastic mouthpiece coated with vinyl-silicone impression material. After a few minutes, the impression material sets, and the mouthpiece is removed. Try getting a monkey to open wide and hold still for the process. You have to sedate them; but anesthesia can cause chewing muscles to clench and saliva to flow. Teaford worked out the details, and by 1989 was ready to take it to the next level, to study tooth wear in wild primates.

Glander began working with wild howling monkeys in 1970 at Hacienda La Pacifica in Costa Rica. La Pacifica is a 2000-hectare farm and cattle ranch a few kilometers down the Pan American Highway from the town of Cañas, in Costa Rica. Glander was interested in howler diets. They eat more leaves than other primates in the Americas. Primatologists at the time believed that leaf-eating monkeys in the trees had an unlimited supply of food. All a howler should have to do is open its mouth, move its head around the canopy, and start chewing. But Glander had a difficult time believing it was so simple. Recall the evolutionary arms race between eater and eaten. Plants have chemical and mechanical defenses to protect their leaves from predators. How do howling monkeys deal with them? Glander spent the better part of the next couple of decades at La Pacifica watching and learning. To this day, no one knows more about what, and how, howling monkeys eat.

Teaford and Glander soon started working together to document the effects of diet on tooth wear in wild primates. They were a perfect match, and La Pacifica was the perfect place. The land was purchased by pharmaceutical giant Ciba-Geigy in the 1950s to grow medicinal plants. But the agricultural engineer sent to manage the place, Werner Hagnauer, quickly realized that the dry season was too long and severe. Instead, Hagnauer set up a cattle ranch, maintaining patches of dry forest as windbreaks to prevent erosion of grazing land. Monkeys lived in those patches. They also lived in a narrow but lush band of forest along the banks of the Río Corabici, a roaring white-water river that cuts through La Pacifica. Not only could Teaford and Glander study tooth wear, they could also compare animals that live in different habitats.

La Pacifica became something of a training ground for a generation of students and postdocs, including me. Glander would dart the monkeys, and we'd catch them in nets as they fell from the trees. We'd bring them back to camp, and Teaford would take dental impressions. Once the anesthesia began to wear off, we'd release them back into the wild. Howlers aren't the brightest of monkeys, and most seem not to mind the process enough to figure out how to evade recapture again and again. Julio and Cindy, Jose and Fiona, Chief and Chaya, Trinka and ET—each year for a decade, Glander, Teaford, and the team returned during rainy and dry seasons to collect dental impressions from these old friends. It was the perfect sample for my study of changes in tooth shape with wear.

I set another graduate student, John Dennis, on the project. And the results confirmed everything we found for the chimpanzees and gorillas. Surface slope and relief dropped with age, but angularity held for years on end. Monkey after monkey, the pattern was the same. It didn't matter whether it lived in the dense forest lining the river or the drier, more open ones separating the grazing lands. The implication was clear. Howling monkey teeth are destined to wear in a specific way. Nature selects not only for tooth shape, but also for an internal structure that guides shape changes through the wear sequence. We could use different individuals, living or fossil, to document the pattern.

FINAL THOUGHTS

Georges Cuvier is reported to have said, "Show me your teeth and I will tell you who you are." It makes sense if teeth are a reflection of diet and, as Cuvier's contemporary Jean Anthelme Brillat-Savarin proposed, you are what you eat. But it also raises some important questions. Biology is a messy business. How tight is the relationship between teeth and diet, really? Think about it. We humans have an incredibly flexible diet, with tastes varying to match. We'll learn in chapter 8 that Gwi San hunter-gatherers of the Kalahari had a traditional diet dominated by melons and roots, whereas Inuit in the High Arctic survived months on end with little other than marine mammals and fish to sustain them. Is this dietary versatility uniquely human? Do other animals have dietary free will, or are they limited by the teeth in their mouths?

This leads us into new territory with very important implications for paleontologists. To get there, though, we need to know not just how teeth work, but also how animals use them. These aren't always the same thing. The form-function relationship in biology is complicated. Understanding this is critical if we are to use the fossil record to explore evolutionary history. Yes, teeth can teach us something about what a species in the past evolved to eat. But can it teach us what an animal in the past actually ate on a daily basis? Sometimes yes, sometimes no. To understand this, we need a better appreciation of what animals eat in their natural habitats, and why.

CHAPTER 2

How Teeth Are Used

"They're not supposed to do that," I told my wife, Diane, as I transcribed field notes by the light of our kerosene lamp at Ketambe, our research station in Indonesia's Gunung Leuser National Park. We had arrived in the rainforest only a week before, and I had just finished following *Antara*, a group of long-tailed macaques that live in a small patch of rainforest along the left bank of the Alas River. Page after page of dietary observations, it was leaves, leaves, and more leaves. But they have the teeth of a fruit eater. Their incisors are large, for peeling husks, and their molars have short crests and rounded cusps for pulping fruit flesh. The common name for their biological subfamily is even "fruit-eating monkeys." Clearly these macaques hadn't read the literature on tooth form and function.

I realized that night that it's one thing to study teeth in the laboratory and another entirely to see how they are actually used by primates in their natural habitat. Both are important. In this chapter we visit primatologists in the field and follow along as they watch their study animals make use of the forest to earn a living. The take-home message is, as I learned during my first weeks at Ketambe, there's more to food choice than teeth. Yes, teeth are important. They give an animal access to foods that would otherwise be off the table. But primates also have to worry about getting the right mix of nutrients to meet their needs,[1] contend with competitors, and avoid predators while feeding. Then there's the issue of availability. Potential foods appear and disappear in the forest like daily specials at the corner deli.

We learned in chapter 1 that primates with blunt molars tend to eat more fruit, and those with long crests eat more leaves. But does that mean food choice necessarily depends on tooth shape? Is the relationship tight enough to use teeth to reconstruct the diets of fossil primates? Studies of animals in the wild give us the context we need to begin to address these questions.

SPECIES-SPECIFIC DIETARY ADAPTATIONS

I spent much of the year before I left for Ketambe casting and measuring the teeth of monkeys and apes in natural history museums throughout the United States and Europe. Many primates were collected during scientific expeditions to northern Sumatra early in the twentieth century. Cabinets at the Academy of Natural Sciences in Philadelphia, for example, are filled with specimens from George Vanderbilt's visit in 1939. My idea was to use the teeth of these specimens to generate hypotheses about tooth use, based on what we know about how teeth work, then test them by observing feeding behaviors of wild individuals of the same species. I assumed, as most others did at the time, that primates seek out and eat the sorts of foods to which their teeth are adapted—those that their species had evolved to eat. Bob Sussman at Washington University in Saint Louis has been championing this idea since the early days of field primatology, developed in part from his work on the Indian Ocean islands of Madagascar and Mauritius.

From Madagascar to Mauritius

Sussman was a graduate student in search of a project at Duke University back in the late 1960s. Those were exciting times in Durham, North Carolina. John Buettner-Janusch had just moved his colony of about 90 captive primates, mostly lemurs, from Yale to a large research sanctuary in the Duke Forest. What was lacking, though, was basic information on the ecology of most lemur species in their natural habitats—how they used the forest to earn a living. Buettner-Janusch encouraged Sussman to go to Madagascar and help discover the details. Sussman first visited the island back in 1969.[2]

Madagascar sits in the Indian Ocean south of the equator, about 250 miles east of the shores of Mozambique. It's one of the largest islands in the world, 1000 miles long and as many as 300 miles wide, and it boasts habitats ranging from desert to grassland to tropical rainforest. It's a magical place for the ecologists who work there. The island has been isolated since the days of the dinosaurs, and there are today thousands of types of plants and animals found only on Madagascar and smaller neighboring islands. Lemurs are among them.

Madagascar is to lemurs as the Galapagos Islands are to finches. Primates arrived there more than 100 million years after it separated from the African continent. Madagascar had no large-bodied terrestrial carnivores or herbivores, as existed on the mainland, as best as we can tell. And there were no monkeys or apes with which ancestral lemurs had to compete. It was a perfect natural experiment in primate evolution, and the lemurs did not disappoint. There are more than 100 species today, with up to a dozen found in any one place.

This created a conundrum for early primatologists. How could two or more species competing for the same resources coexist? According to evolutionary theory, the best-adapted one should have outcompeted the others, forcing less "fit" species to turn to alternative resources or face extinction. How could so many closely related, similar primate species live together in the same place at the same time? Sussman wanted the answer, and Madagascar was a great place to find it.

Alison Jolly, who had studied Buettner-Janusch's captive primates when they were still at Yale, was among the first to document the ecology of lemurs on Madagascar. She began her work in the Berenty Private Reserve at the south end of the island in 1963, and had already published an influential book, *Lemur Behavior: A Madagascar Field Study*,[3] when Sussman started. In it she described a group of ring-tailed lemurs that had adopted a solitary brown lemur. Brown lemurs were not endemic to Berenty. This one had been captured about 50 miles away, was brought to the area, and then escaped from captivity to join his ring-tailed cousins. The brown lemur slept with the ringtails, moved with them, ate with them. Did this mean that the two species might use the same resources where their ranges overlapped? No one had yet documented the diets of brown lemurs in the wild in detail, let alone the effects of competition on coexistence with ringtails. Sussman seized the opportunity to add to the growing body of knowledge on lemur ecology and, at the same time, learn how two closely related primate species of similar size and shape might divvy up forest resources to live together in peace.

Bob Sussman and his wife Linda headed to the uncharted lowland forest in the southwest to establish their own research camp. They found two forest patches—Tongobato, which had brown lemurs but no ringtails, and Antserananomby, which had both. The plan came together. He would study brown lemurs at both sites and compare their diets to determine the effects of competition with ringtails. He would also study ringtails at Berenty, which, again, had no groups of brown lemurs, to see how they differed there. The three forests were very similar, especially Tongobato and Antserananomby, which were separated by just a few miles of cultivated fields and degraded wild vegetation. Berenty had a few species of large trees not found to the west, but the forest canopies at all three sites were dominated by tamarind trees. These trees produce podlike fruit used in many cuisines throughout the world today. You might know tamarind from the distinctive flavor of Worcestershire sauce. Lemurs also eat tamarind fruits, along with their leaves, flowers, and bark, depending on the season. The study design was completed by planning to work at Antserananomby in the dry season and Tongobato and Berenty in the rainy season. That way, Sussman could compare diets of each species between seasons as well as between sites where they lived together and separately.

After months of study in 1969 and 1970, he had his results. The two species did overlap in their diets—both ate fruits, leaves, flowers, and bark. But the brown lemurs concentrated more on leaves, and ringtails consumed more fruits and ate a greater number of plant species.[4] The ringtails moved about half a mile each day and ranged in height from the tops of the trees to the ground. The brown lemurs, in contrast, moved only about 100 yards and stayed in the main canopy. These differences made sense in terms of their diets and were consistent between sites and between seasons, whether or not the same exact foods were available at a given time and place. In other words, the lemurs sought out and ate the variety of foods to which they were adapted regardless of where, when, or with whom they were eating. This allowed them to split the forest's resources where they coexisted.

But what about primates living in radically different environments? Would they still seek out and eat particular types of food? Sussman followed up his lemur study with macaque monkeys on the idyllic tropical

2.1. Ring-tailed lemurs live and eat in a variety of ecological settings. Photographs courtesy of Robert Sussman.

island of Mauritius, about 450 miles to the east of Madagascar. He hadn't intended to work there. He was actually en route to Madagascar in 1977 when his flight from Paris to the capital city, Antananarivo, was cancelled. Sussman, along with his wife and his colleague Ian Tattersall, an anthropologist from the American Museum of Natural History, managed to get rerouted to Mauritius in the hopes that they could catch a flight on to Antananarivo from there. But Madagascar was still reeling from the assassination of President Richard Ratsimandrava and effects of the military coup that had established a new socialist government. Demonstrations erupted in Antananarivo because of shortages of food and other commodities. There was simply no way to get there.

Could they somehow salvage the trip? Mauritius had no lemurs or other endemic primates to study, but there was one invasive species of monkey, the long-tailed macaque. Macaques had been brought to the island from Indonesia by European sailors around 450 years ago, and their population had exploded to tens of thousands. This could be the start of an interesting project. Sussman spent that summer surveying macaque populations and returned a couple of years later to compare the diets of groups living in degraded savanna habitat with those in undisturbed forest.

Just like the brown lemurs and ringtails, macaque populations living in different sites had similar diets, both in terms of diversity of plant species and proportions of different plant parts eaten. In this case, though, the primates lived in very different settings, with access to very different sorts of foods. In fact, the Mauritius macaques even had similar diets to those living thousands of miles away in their native forests of Southeast Asia.

This study, together with his earlier work in Madagascar, led Sussman to conclude that primates have what he called *species-specific dietary adaptations*. Their food preferences depend on the teeth and guts that gave their ancestors an advantage in acquiring and processing the sorts of things they had evolved to eat. This is why different primate species living together had different diets despite access to the same foods. This is why populations of the same species living in different settings sought out similar foods despite having access to different resources. To Sussman, diet was related, first and foremost, to adaptations of the teeth and digestive tract inherited from one's ancestors. It was a matter of evolutionary legacy.

Sussman's results were like music to a paleontologist's ears. They gave extra credibility to reconstructions of diet based on fossil teeth. We can't infer all the details of the foods available to past primates, or exactly how competitors forced them to share the resources of their forests. But if Sussman's lemurs and macaques were anything to go by, teeth should give a pretty good indication of diet regardless. This also made sense in light of Rich Kay's studies of molar shearing crest length and crushing area (see chapter 1). Remember that primates classified as leaf eaters tend to have longer crests and smaller crushing areas than do fruit eaters.

Nature Is Anything but Simple

That said, it is one thing to demonstrate that living primates with a given tooth shape tend to eat specific types of food, but quite another to be confident that the form-function relationship is strong enough to be used to reconstruct the diet of a given fossil species. As Richard Preston wrote in *The Hot Zone*,[5] "In biology, nothing is clear, everything is too complicated, everything is a mess, and just when you think you understand something, you peel off a layer and find deeper complications beneath." To be sure, folivorous primates tend to have longer shearing crests than do closely related frugivores today. But how do we know that the connection is sufficiently tight to use crest length to determine whether a fossil species ate fruits or leaves?

One approach involves pretending specific living primates are fossils, inferring their diets based on tooth shape, and seeing how closely the "reconstructed" diets match actual ones. We can look, for example, at brown lemurs and ringtails. Rich Kay, Bob Sussman, and Ian Tattersall did just that back in the late 1970s. Surprisingly, the two species have remarkably similar teeth. If we found them in the fossil record, we'd infer that they had the same diet. I guess they just don't realize that they should eat the same foods. Brown lemurs consume more leaves and ringtails take more fruit, despite the fact that their shearing crests are the same length.

So the relationships between teeth and diet are not so simple after all. Truth be told, the two lemur species do eat mostly the same sorts of foods—fruits, leaves, flowers, and bark—just in different proportions. Kay, Sussman, and Tattersall speculated that perhaps tooth shape is more about the *types* of foods a species eats than the frequencies in which they are eaten. In other words, maybe shearing crest length doesn't reflect how many leaves you eat but, rather, whether you need to be able to eat them at all. Kay, Sussman, and Tattersall evidently didn't realize it at the time, but they had hit on a profound new idea. It would take a couple of decades, though, and work on a very different kind of primate, to bring it to the forefront.

GORILLAS IN OUR MIDST

Gorillas may not seem like an obvious choice if our ultimate goal is to understand what teeth can teach us about the evolution of human diet. They have highly specialized sharp molars with long crests and thin tooth enamel. These are almost the antithesis of early hominin teeth, which are flat and covered in an extra-thick coating of enamel. On the other hand, gorillas are among our closest living relatives, and have been studied in the wild longer than any other great ape. We know a lot about their food preferences in time and space and across seasons and sites ranging from high mountaintop to lowland forest. If we can't work out tooth form-function relationships for gorillas, we have no hope of using teeth to reconstruct the diets of our distant ancestors and other fossil primates.

The Virungas

We can start off, like the first field researchers to study gorillas, in the majestic chain of volcanoes straddling Uganda, the Democratic Republic of Congo, and Rwanda. This is central Africa's Virunga range, its cloud forests shrouded in mist and topped by peaks towering from 10,000 to nearly 15,000 feet above sea level. The Virunga Mountains are home to a small, isolated population of gorillas studied with only brief interruptions for more than half a century.

George Schaller was the first in. He was still a graduate student at the University of Wisconsin at the time. He and his wife, Kay, established their base in 1959 at Kabara, between Mount Mikeno and Mount Karisimbi, in what is today the Democratic Republic of Congo. That pioneering work inspired a generation of researchers, including the famed primatologist Dian Fossey. Fossey had in fact intended to work at Kabara herself in 1967, but was forced to move camp to the other side of Mount Karisimbi, to Karisoke in Rwanda, because of political upheaval and the turmoil of the Congo Crisis. Conditions haven't always been great at Karisoke either, but the research center there has survived war and looting, with only brief interruptions of work since Fossey first established it. It has given us an amazing window into the life of the mountain gorilla.

You can read Schaller's and Fossey's remarkable stories in their books, *Year of the Gorilla* and *Gorillas in the Mist*, respectively.[6] The Virunga gorillas were the first that scientists really came to know in the wild, the first to share with us the intimate details of their lives. Back when I was in graduate school, almost everything we knew about gorilla ecology we had learned from them. Schaller, Fossey, and the primatologists who followed discovered not the brutish, murderous beasts imagined by nineteenth-century naturalists and portrayed by early Hollywood, but intelligent, peaceful plant eaters.

The Virunga gorillas live in montane forest and woodland habitats that average two miles in altitude. They regularly climb to more than 12,000 feet above sea level, where mist sets in, temperature drops, and wind speed swells. Mountain gorillas there eat mostly stems, leaves, and pith of nonwoody plants, like wild celery, on or near the ground. These foods are called terrestrial herbaceous vegetation (THV for short). THV is plentiful and available all year round. Mountain gorillas also have a penchant for bamboo shoots; they travel to the lower reaches of their ranges when these tender plant parts are in season.

This diet makes a lot of sense. Adult males typically weigh 350 pounds. They need a lot of food, and the forest is filled with THV. Gorillas also have massive guts that house countless microorganisms to break down otherwise indigestible, fiber-rich plant parts. They spend a long time digesting, about 60% longer than chimpanzees, to squeeze as much energy as possible from foods reluctant to give up their nutrients. More to the point, they have big, sharp molar teeth, with long crests for shearing and slicing leaves and stems, and strong, deep jaws to withstand the beating that comes from the incessant chewing needed to puree these tough foods.

This anatomy-to-diet relationship seems even clearer when we contrast the gorilla with the other great apes, the chimpanzee and the orangutan. Chimpanzees specialize on soft, fleshy fruits and orangutans eat more hard foods, like bark and shell-covered nuts. Both have large incisors for husking fruit peels and rounded, bowl-shaped molars for pulping fruit flesh. Orangutans also have thickened enamel covering their teeth, which strengthens and protects them from breaking given the concentrated force needed to crack open hard shells. Gorillas and chimpanzees have much thinner tooth enamel. Nature seems to give us three very clear examples of relationships between teeth and diet in our closest living relatives—or so we thought when I was in graduate school.

Most researchers now recognize two species of gorilla, one in the east (*Gorilla beringei*), and the other in the west (*Gorilla gorilla*). The mountain gorilla belongs to the eastern species. We know the most about them because George Schaller chose to work in the Virungas. The gorillas there were on the verge of extinction, so there was a sense of urgency to document their lives before it was too late. They were also easy to find because American and European expeditions had visited the area and collected specimens earlier in the century. Dian Fossey followed, and the rest is history. When we think of gorillas today, most of us envision them munching on wild celery in the mist. But that's a historical accident. The Virunga mountain gorillas are not typical. They are a peripheral, marginal population, living in the most extreme habitat of any ape species. They are the rarest of the gorillas too. Just a few hundred of them survive today.

Bai Hokou

On the other hand, there are 200,000 gorillas 1000 miles to the west in the lowland rainforests of the Congo Basin. Researchers began chasing them in the 1960s, but most early studies of their diets relied on fleeting glimpses, searching for dropped foods along fresh trails, and sorting through feces left behind. Western lowland gorillas were more difficult to study than mountain ones because they were more skittish. They spent more time in the trees hidden behind dense vegetation, and their groups were more apt to spread out. It was obvious from the start, though, that gorillas in the western lowlands ate more fruit than did their cousins in the Virunga Mountains. Still, the details remained sketchy. It wasn't until the 1990s that western lowland gorillas began to feel comfortable enough with primatologists to offer them the level of detail researchers had come to expect from studies in the Virungas.

Melissa Remis was among the first invited in. She worked in the Dzanga-Ndoki National Park, a thin sliver of southwestern Central African Republic wedged between Cameroon and Congo. The park is mostly pristine lowland rainforest broken by natural clearings called *bais* by the BaAka pygmies who live in the area. Remis was a graduate student at Yale in the early 1990s and focused her work on gorillas at Bai Hokou around the same time that I was at Ketambe. The Bai Hokou gorillas were still wary, but they occasionally let her watch from a distance, at least when they were in the trees. Remis was only able to clock a couple of hundred hours of direct observation in two years, but it was enough, when combined with studies of feces and dropped food, for the most complete picture of western lowland gorilla diets at the time.

To make a long story short, the gorillas at Bai Hokou ate fleshy fruits with succulent, pulpy insides whenever and wherever they could find them. They would sometimes walk half a mile or more, right past edible leaves and stems, to get to a fruiting tree. The gorillas worked hard to get fruit. That's not to say that they did not eat the more fibrous THV that mountain gorillas consumed so much of. They did. It's just that they seemed to prefer sugary fruit flesh when it was available. In fact, availability is the key to understanding the diet of Bai Hokou gorillas.

The Dzanga-Ndoki National Park has a very distinct dry season. During the three months between December 1991 and February 1992, for example, it rained less than an inch despite an average annual rainfall of more than 55 inches. Plants tend to flower late in the dry season, leaves sprout, or flush, with the onset of the rains in March, and fruit production peaks in the middle of the rainy season, around July or August. Unsurprisingly, gorillas eat more fleshy fruits than anything else during much of the rainy season. Other foods, such as tree leaves, stems, and bark, and THV are eaten year-round, but they only dominate the diet during the dry season.

So, the western lowland gorillas at Bai Hokou were, at least during the rainy season, fruit eaters. This should not have come as a surprise. The nineteenth-century anatomist Richard Owen gathered what information he could from travelers and residents of the coastal lowlands of tropical West Africa, where "fruit trees of various kinds abound both on the hills and in the valley." These, according to Owen, "afford the great denizen of the woods a successive and unfailing supply of indigenous fruits."[7] But the idea of a frugivorous gorilla still did not sit well with everyone. Some reviewers had a hard time with Remis's results because decades of work in the Virungas had taught us that gorillas eat leaves, wild celery stalks, and other tough herb parts on the ground. Could animals with guts and teeth so well evolved for low-quality, fibrous foods actually *prefer* to eat fleshy fruits? Could gorillas really be seasonal frugivores? Or was it simply that observations at Bai Hokou were somehow biased because the animals were skittish and more difficult to see and follow than in the Virungas?

The San Francisco Zoo

Remis needed evidence that gorillas prefer sugary, fleshy fruits to tough, fibrous foods. It made intuitive sense. If you give a five-year-old child the choice between a chocolate bar and a head of broccoli, the

2.2. Zoo gorilla feeding.

chances are the kid will pick the chocolate bar. Wouldn't a gorilla too? While chocolate isn't an option, we might still expect gorillas to choose sugary, fleshy fruits over celery or cabbage when given a choice. Remis spent a couple of months in 1999 with a half dozen gorillas at the San Francisco Zoo to find out.[8] A typical captive gorilla eats eight or nine pounds of mixed fruits and vegetables and one to two pounds of leaf-eater primate chow each day. The chow is nutritious and comes in biscuits about the size of a small cookie, though I'm told they don't taste much like cookies. Gorillas at the San Francisco Zoo were fed in two ways. They ate morning and evening meals by themselves in their cages to make sure everyone had their fair share, and had midday snacks together as a group in their outdoor enclosure as part of the zoo's enrichment program. This gave Remis the chance to look at food preferences of the gorillas in both individual and group settings.

How do you ask a gorilla what it likes to eat? You do paired-feeding experiments. Put two different foods in front of one and see which it takes first. Would it take the one higher in sugar? What about protein or fat? Would it avoid plants with chemical defenses, like toxins or tannins that leave a bitter or astringent aftertaste? What about plant parts with physical defenses, like tough, fibrous foods, or ones with hard shells? Remis scoured the Asian and Hispanic food markets of San Francisco for fresh fruits and vegetables as close to those available to wild gorillas as she could find. There were limits—not many toxic fruits or leaves in food co-ops. But in the end, she was still able to give the gorillas lots of options: sweet mango, bitter tamarind, sour lemon, tough celery, and many others.

Just before mealtimes, Remis put two different foods on opposite sides of each cage and watched to see which the gorillas took first. They were especially excited to get mangos. One male even gave copulation calls when she arrived with a bushel! Figs and bananas were also met by pig-grunts of approval if not delight. Kumquats, parsnips, lemons—not so much. The midday snacks were telling too. The dominant animals scooped up favored foods hurriedly, like a child grabbing as many chips as she can hold before a sibling can get a hand into the bag. Mango, cantaloupe, and corn went especially quickly. Outranked gorillas were left with broccoli, cabbage, kale—and celery. The results were clear. Gorillas preferred foods high in sugar and low in fiber when given a choice, just as most people do.

But there is another possible explanation for the difference between Remis's gorillas and those from the Virungas. The gorillas in the San Francisco Zoo are, like those at Bai Hokou, the western lowland variety. In fact, nearly all captive gorillas are. Can we assume that the mountain gorilla, a different species entirely, has the same preferences? Would the Virunga celery munchers also prefer fruit if given the choice? Or do western lowland gorillas simply have more of a sweet tooth than their mountain gorilla cousins? The short answer to this last question is, probably not.

Uganda's Bwindi Impenetrable National Park also has mountain gorillas, no different genetically than those from the Virungas, which are 20 miles to the south across clear-cut, cultivated hills. The Bwindi ones live at lower altitudes though, between about 5000 and 7500 feet above sea level. Bwindi also has a greater diversity and density of fruit; unsurprisingly, the mountain gorillas there eat more of it. The same goes for the Grauer's gorillas found in the lowlands and moderate elevations of the eastern Democratic Republic of Congo and along the border with Uganda and Rwanda. These too are the same species as the mountain gorillas, and they eat fruit in proportion to its availability throughout the year. In fact, there is a strong correlation between altitude, fruit availability, and fruit consumption across gorilla ranges whether you're considering eastern or western gorillas.

But this brings up another question. Given that gorillas prefer fruit, why do they live in the Virungas at all? It is cold and wet, and there's little other than wild celery to eat much of the year. The answer may be that they have nowhere else to go. The Virunga Mountain parks and Bwindi are islands isolated by expanding human settlement. The nutrient-rich soil of the region supports one of the densest rural human populations on the African continent. While western lowland gorillas have access to fruit much of the year, hunting and forest clearing for cultivation and firewood have pushed the Virunga gorillas to altitudes where they have no choice but to fall back on THV. They do it not because they want to, but because they have to. They do it because their anatomy allows them to. We might call them perpetual fallback feeders.

In sum, gorillas eat soft, sugary fruit flesh when they can get it, but have the tooth and gut adaptations needed to make do with tough, fibrous vegetation when they can't. In this case, teeth and guts give gorillas the option to eat less preferred foods. Teeth are about accessibility. Food choice, on the other hand, is also about availability. Gorillas teach us that these are not necessarily the same things. If our ultimate goal is to understand the evolution of diet, we need a handle on both. We need to know not just what animals *can* eat, but what their options are. We need to understand how availability changes. The place I know best how that works is Ketambe.

THE GUNUNG LEUSER NATIONAL PARK: A TIGHTLY WOVEN WEB OF LIFE

The Gunung Leuser National Park is part of the Tropical Rainforest Heritage of Sumatra, a UNESCO World Heritage Site. The site straddles the equator along the main spine of the Bukit Barisan mountain range, which runs 1000 miles down the western side of the island. The Ketambe Research Station occupies a small part of this expanse, a couple of hundred hectares of virgin rainforest bounded in the north by

two rocky rivers, the Alas and the Ketambe, and in the south by mountainous escarpments of the Barisan range. Still, there is a seemingly
endless variety of plants and animals there, and researchers have been
studying the seven primate species for more than four decades. My wife,
Diane, and I spent thousands of hours with those monkeys and apes
back in the early 1990s.

The first thing that strikes you when you begin a primatology field
project is that your study animals are merely a small part of a very big
forest. It overwhelms the senses—the innumerable shades of brown
and green reaching to the sky, chatter of birds and cicadas, fragrance
of flowers and earth, and sting of nettles and ants. We were surrounded
by—no, part of—a web of life so tightly woven that the rainforest felt
like one living, breathing being with all of its parts working together in
sync. Lesson number one: monkeys and apes are part of a larger whole,
and you can no more understand them outside of this context than you
could understand how a brain or heart functions after cutting it out of
the body.

The Rhythm of Life

The rainforest thrives when its plants and animals work together. A fig
tree produces fruit. A monkey eats it and disperses the seeds with a nice,
neat package of fertilizer. A dung beetle buries the seeds, and a new
plant emerges and begins to grow. It is an incredibly complex arrange-
ment with interactions between the parts constantly shifting to keep
the system in balance. And timing is crucial.

This timing is the focus of phenology, the study of plant and animal
life cycles and what controls them. Where I live on the Ozark Plateau
in northwest Arkansas, April showers bring May flowers; leaves flush
in March and fall in November; blackberries grow in July; and apples
ripen in October. These are part of the normal life cycle of plants, and
they come like clockwork most years, tied to annual changes in temperature, rainfall, and day length. Advances in understanding how climate change between years affects flowering, flushing, and fruiting are
coming fast and furious today. Global warming can alter or disrupt life
cycles, and phenology plays an important role in helping us understand
the consequences. But that's another story.

Phenology is important to primates and their teeth because plant life cycles determine the foods available in a given place at a given time. The rainforest is like the produce section of your local supermarket, with some seasonal items and others available year-round. Primates can pick and choose, depending on the costs and benefits of individual foods, what to fill their carts with from the items in stock at the time.

While it is clear that the seasons affect plants and animals in higher-latitude places like the Ozark backwoods, what about in the tropics, where most primates live? Most tropical plants produce new leaves, flowers, and fruits in bursts. These bursts can be synchronized within or among species of plant in a forest, or they can be spread throughout the year. And whether a plant community staggers or clumps its life-cycle events has important consequences for herbivores. It affects the number of animals a forest can support, the strategies they use to forage for food, and the timing of their own life-cycle events, such as seasonal breeding and birthing intervals.

Phenology can be complicated because there are many factors that influence when plants fruit, flower, and grow leaves. We must first distinguish between *proximate* causes, those that trigger individual events, and *ultimate* ones, the evolutionary reasons for their timing.

The common triggers for life-cycle events, such as flowering, flushing, and fruiting, are annual changes in rainfall, temperature, and solar energy.[9] Some sites vary more in climate over the year than others, and life-cycle events tend to be better synchronized where there are distinct seasons. Proper timing can help plants deal with regular and recurring environmental stressors such as annual droughts or limited sunlight. A tree might bear fruit at the beginning of the rainy season so seeds have the moisture they need to germinate quickly, to minimize the time they lie dormant on the forest floor. Leaves may sprout and grow as sunlight peaks because younger ones are better at tapping solar energy to make food from carbon dioxide and water. But different plants play by different rules. Larger trees that store more carbon and water can have longer fruiting seasons spread throughout the year, and fruiting peaks in the canopy and understory can be offset because conditions vary with height. More important, each forest has a unique soil and drainage pattern, topographic features, and biological community, all of which can affect plant parts available for a primate to eat at any given time. Still, most of the places we find primates have some clumping and some stag-gering, largely because their biological communities are complex.

This is where we get into ultimate causes.[10] Plant communities some-times have fruit production timed so that birds passing through during annual migrations can spread their seeds. Clumped flowering might provide the abundance needed to attract pollinators. Synchronizing events can also ease the effects of damage caused by predators. It would take a lot of squirrels to eat all the acorns produced by a stand of oak trees during the autumn burst here in the Ozarks. Also, flushes might be timed so that young, vulnerable leaves emerge when the insects that eat them are scarce. On the other hand, staggering fruiting, flowering, and leaf production can also be advantageous. It can minimize compe-tition between plants trying to attract fruit dispersers or pollinators, and can prevent the critical mass of food needed to support seed or leaf predators in the forest. Whatever the circumstances of a particular for-est, the timing of life-cycle events helps the ecosystem keep its delicate balance. These are the sorts of things primates need to be aware of as they plan their menus for the year.

So what about Ketambe? The research area averages more than 10 feet of rain in a typical year, peaking at about 16 inches a month around April and November. There are two rainy seasons and two less-rainy ones. You can't really call the less-rainy ones dry seasons though, because even the driest months typically have six or seven inches of rainfall. But you can certainly tell the difference. During the rainy sea-son things are always wet and moldy. You spend most of the day hud-dled under a poncho trying to keep the leeches out and your binoculars from fogging over. In July and January, on the other hand, there are more sunny days than rainy ones, and your clothes and towels can dry on the line the same day they're washed.

Forest phenology reflects the seasonality at Ketambe. Primatolo-gist Carel van Schaik and his field team measured fruiting, flowering, and leaf flush over the course of several years back in the 1980s. Each month they walked many miles through the rainforest, checking hun-dreds of trees for new growth, counting fruits that fell on the trail, and picking litterfall from nylon nets. They found that immature leaves were most abundant from December to February (the first less-rainy season), flowers between January and April (first less-rainy and rainy seasons), strangler figs in April and October (wet-season months), and other ripe fruits in July and August (second less-rainy season).

This pattern held while we were at Ketambe too, but the rules are somewhat different in some years, especially for the majestic dipterocarp trees that tower over the canopy. Every few years rainfall levels drop, following roughly a cycle corresponding to the El Niño–La Niña Southern Oscillation. For reasons we don't fully understand, the sea-surface temperature of the tropical eastern Pacific Ocean becomes especially warm or cool every few years. These events, called El Niño and La Niña, respectively, affect weather on a global scale. Fluctuations in sea-surface temperature and air pressure alter wind patterns and ocean circulation, which in turn impact the distribution of rainfall across the globe. In Southeast Asia, droughts commonly occur during the transition period from La Niña to El Niño. Especially dry periods in turn trigger heavy flowering and fruiting, doubling production over other years. So there's more to cycles of food availability in the forest than mere annual seasonality.

These all affect primate food choice. It reminds me of dinner at home when my kids were young. I remember one evening my younger daughter wasn't happy with what she'd been served, and her sister responded sternly, "You git what you git and you don't throw a fit." That's the way it works in the real world. Primates at Ketambe and elsewhere are limited to the food available at the time. If they don't like it, it's too bad. Whether or not they have long shearing crests, the choices may be leaves or nothing. So to learn why primates eat what they do, we need to know what's on the menu and understand how it changes over time.

Teeth Matter

That is not to say that teeth don't matter. They do. And Ketambe is a great place to examine relationships between teeth and diet under real-world conditions. There are seven species of primate living in the same patch of forest with access to the same foods at the same time. But they have different tooth and gut specializations to work with, and these clearly affect food choice. Just as on Madagascar, different primate species might eat the same types of fruit, leaf, or insect at any given time, but differences in food preference emerge over the course of a year. The leaf monkeys at Ketambe eat more leaves and seeds. Macaques, gibbons, and orangutans eat more fleshy fruits. Gibbons seek out small, berry-sized fruits, whereas orangutans prefer larger ones. And macaques eat more insects than do the other primates. These differences over time are about what we'd expect given variation in the sizes of the primates, and the shapes of their teeth and guts. They are the sorts of differences that over the long term allow primate species to coexist and share the resources in a forest.

But it's not just these broad differences in diet that matter. Sometimes it's the subtle ones that best show how teeth help primates decide who eats what in the forest. One example stands out in my mind. The fruit called *akar palo* in the local language (identified as *Gnetum* cf. *latifolium*) is about the size of a plum, but it has a hard shell and pulpy innards surrounding several large, soft seeds. Gibbons, macaques, and leaf monkeys all break through the husk with their incisor teeth and pry pieces off to expose the edible parts. It's a slow and tedious process; it takes more than a minute for them to eat a single fruit. Orangutans, on the other hand, crush *akar palo* fruits between opposing teeth to split them open. They can eat several in a minute, and this gives them an advantage.

But there's more. *Akar palo* fruits harden as they ripen, and the gibbons, macaques, and leaf monkeys eat only the softer, unripe ones. I saw several of these primates try, unsuccessfully, to pierce the husk of a ripening fruit, then drop it uneaten. The orangutans continue to visit *akar palo* vines and eat their fruits days and sometimes weeks after the other primates in the forest are forced to abandon them. Strong, thickly enameled teeth and powerful jaws give the orangutans an edge over the other primates in the forest.

While orangutans eat hard foods that other primates at Ketambe tend to avoid, like ripe *akar palo* fruits, acorns, and bark, they still consume a lot of the same soft, fleshy fruits as the macaques and gibbons there, and the same young, succulent leaves as the leaf monkeys. Orangutans are no more hard-object specialists than western lowland gorillas are leaf and stem specialists. Both the orangutan and gorilla have dental specializations to overcome mechanical defenses of challenging plant foods. Both eat foods that match their teeth. But that doesn't mean they eat them all of the time, or even most of it.

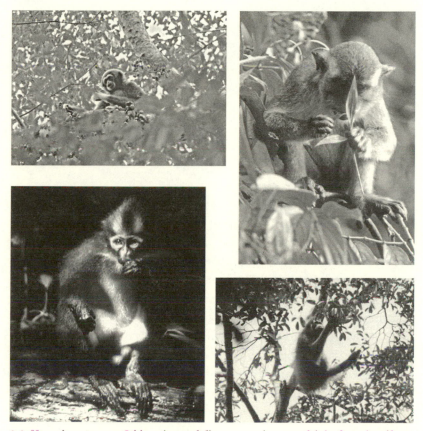

2.3. Ketambe primates. Gibbon (*upper left*), macaque (*upper right*), leaf monkey (*lower left*), and orangutan (*lower right*). All live in the same patch of forest.

This brings us back to the brown lemurs and ringtails, and their similar teeth despite their dissimilar diets. Recall that they both eat fruits, leaves, and bark, but in very different proportions, and also that Rich Kay, Bob Sussman, and Ian Tattersall suggested that the types of foods lemurs eat might be more important for nature's choice of tooth shape than how often each is eaten. This certainly seems to be the case for the orangutans at Ketambe, and the gorillas at Bai Hokou for that matter. By analogy, it doesn't matter what kind of teeth you have if you eat pudding 11 months a year. But if you need to eat rocks the other month or you'll starve, your teeth had *better* be adapted for rock eating.

MANGABEYS IN HARD TIMES AND GOOD

Maybe nature doesn't care so much whether a primate takes advantage of its specialized teeth every day or just on occasion, so long as it allows individuals to survive and reproduce. There is no better example of this than a comparison of mangabey monkeys in Kibale National Park of Uganda with those in Taï National Park of Côte d'Ivoire. Let's head back to Africa.

Kibale National Park

Kibale National Park is northeast of the Virunga volcanoes. It's just up the western shore of Lake Edward and over the Rwenzori Mountains. The park includes nearly 300 square miles of protected rainforest on the edge of Africa's Great Rift Valley in southwestern Uganda. Kibale is home to a dozen species of primate, and it has one of the densest concentrations of monkeys and apes in the world. Much of the research on these animals takes place near the village of Kanyawara, in the north of the park. Joanna Lambert, a primatologist at the University of Colorado, has worked there for the better part of the past quarter century. This gives her a marvelous perspective on how food availability varies over the longer term at Kibale and how this affects the diets of primates there.

Lambert was there during the summer of 1997 to study cheek-pouch use in two species of monkeys—red-tailed guenons and grey-cheeked mangabeys. These monkeys have special pouches in their cheeks that allow them to collect and store food, and she wanted to better understand how and why these are used. But soon after Lambert arrived, something else caught her attention. The forest was dying of thirst. The leaves of the understory were wilted and fruit was scarce. Drought had come to Kibale, and its effects were worse than Lambert had ever seen before or has seen since.

Variation in rainfall between years is common for southwestern Uganda. It coincides with the El Niño–La Niña Southern Oscillation, just as drought follows the oscillation at Ketambe. In southwestern Uganda a complex landscape of mountains and lakes adds to the effect. There was an especially strong El Niño event in 1997, and the forest

2.4. Sooty mangabey eating a *Sacoglottis* nut on the forest floor in the Taï National Park (*left*), and a grey-cheeked mangabey eating a soft fruit in the Kibale National Park (*right*). Photographs, (*left*), courtesy of Scott McGraw, and (*right*), Alain Houle.

at Kibale was reeling. The monkeys were hungry. They woke up early, spent most of the day foraging, and often ate things, like tough old leaves, that they would not normally eat. The guenons and mangabeys at Kibale usually eat pretty much the same sorts of foods: soft, fleshy fruits and young leaves. But these were few and far between that summer. Lambert began to notice differences in their diets—the mangabeys were eating more bark and hard seeds, but the guenons were not.

While she set out to study cheek-pouch use that summer, Joanna Lambert discovered something much more important to those of us interested in the relationship between teeth and diet. Her observations at Kibale teach us that while two primate species may have similar diets, even with annual changes in food availability, differences can still surface with extreme conditions brought on by rare climatic events, such as a particularly strong El Niño. Fallback adaptations, those that allow a primate to eat less-preferred foods when favored ones are unavailable, can operate at many different scales. Lean times can come with the ebb and flow of the seasons, but they can also vary across years. In this case, it took a once-in-a-generation drought for diet differences to kick in. Lambert's mangabeys were able to shift to bark and hard seeds because they have anatomical specializations for hard-object feeding. They have thickly enameled, strong teeth and big, heavy jaws well suited to crushing hard, brittle foods. Specialized anatomy gives mangabeys at Kibale an edge too, but there, they only take advantage of it once in a generation.

Taï National Park

It's a completely different story for the mangabeys at Taï National Park. Taï is in the southwest corner of Côte d'Ivoire, 2600 miles to the west of Kibale across the Congo Basin and Gulf of Guinea. It's the largest remaining tract of pristine tropical rainforest in West Africa, and home to 11 primate species. The density of primates there is lower than at Kibale, but the national park itself is four times the size, and the diversity of its species is actually quite similar. Both parks have chimpanzees, two or three types of lower primate, and eight monkeys, including mangabeys.

The mangabeys at Taï and Kibale are very different in some ways. The grey-cheeked species at Kibale prefers the trees, whereas the sooty species at Taï moves and forages mostly on the ground. Back when I was in graduate school, most researchers thought these monkeys belonged to a single group. They both have large incisors; flat molars with thickened enamel; and strong, heavy jaws for crushing hard foods like nuts and bark. Studies of their genes have since made it clear, though, that the mangabeys are not a natural group. In fact, the grey-cheeked species is actually more closely related to savanna baboons than it is to the sooty species. The differences are subtle, but we can see them when we scrutinize the details of their anatomies, including the teeth. This makes the overall resemblances, converging on a specialization for hard-object feeding, all the more remarkable.

Scott McGraw and I were graduate students together at Stony Brook University in the early 1990s, though he started a few years after me. He began working on his dissertation research at Taï in 1993, the year after I graduated. He set out, at least at first, to understand how the monkeys there split up the forest's resources so they could coexist; much as Bob Sussman, Joanna Lambert, and I had all done at our sites. It was the thing to do at the time. Unlike us, though, McGraw focused on differences in how and where they positioned themselves and moved, rather than on diet per se. It was obvious from the start that the mangabeys at Taï were fundamentally different from the other monkeys there. They spent most of their time on the ground, foraging for fallen nuts in the leaf litter. And while his initial observations were not focused on feeding, it was hard for McGraw to ignore it. A group of 100 mangabeys gathered together to crack open nuts on the forest floor can make quite a racket.

McGraw continued to work at Taï on and off over the next few years, and his attention was drawn more and more to the nuts that litter the forest floor. Giant *Sacoglottis* trees typically rise more than 120 feet. They dominate the canopy at Taï and produce huge bursts of fruit for a couple of months each year. The ripe, fibrous pulp smells a bit like apple pie, and it feeds many of the primates in the forest. The fruits also have rot-resistant pits, about the size and shape of a walnut, which can be found year-round on the forest floor beneath the trees. The pits have an extremely hard outer layer, or endocarp, which encases softer seeds. The mangabeys at Taï spend more than half their annual feeding time eating them.

Like the mangabeys at Kibale, those at Taï have flat, thickly enameled back teeth; strong jaws; and powerful chewing muscles—and McGraw had found that sooty mangabeys use their premolars and molars to crack the hardened endocarp. Other monkeys don't eat *Sacoglottis* seeds because they can't. Chimpanzees eat them on occasion, but only by breaking the endocarp with stones rather than teeth. Their teeth and jaws give the mangabeys at Taï an advantage—year-round access to a resource that the other monkeys simply cannot use. This is very different from the situation at Kibale, where the advantage is for consumption of fallback foods eaten only during crunch times when more preferred ones are not available.

The mangabeys and guenons at Kibale have diets much more similar to each other than do those at Taï. Admittedly, these are different species—but, still, why don't sooty mangabeys eat more sugary fruit flesh, like other self-respecting fruit-eating monkeys? Would they pass over *Sacoglottis* nuts, as the Bai Hokou gorillas do edible leaves and stems, if they knew of ripe fruit in the canopy nearby? Maybe they would, but the question is moot. As McGraw learned, the sooty mangabeys at Taï spend most of their waking hours on or near the ground.

If they don't forage in the trees, canopy fruits are just not available to them, whether or not those fruits are ripe. Perhaps the terrestrial-feeding niche is limiting, but it's also what defines these monkeys and allows them to coexist in peace with the other primates in their forest. Again, the relationship between diet and teeth can be very complex, and it can vary depending on the primate species being considered and the resources it has access to.

BELLYING UP TO THE BIOSPHERIC BUFFET

As I mentioned in the introduction, when I think about food choice in nature, I envision the biosphere, that part of our planet that harbors life, as an enormous buffet. Animals belly up to the sneeze guard with plate in hand. What do they take? That depends in part on the utensils they've got to eat with, but also on what's left when they get to the front of the line. In other words, food choice is a matter of matching needs with items that teeth make accessible and nature makes available in a given place at a given time.

All of the primates we've met in this chapter teach us that the right teeth can be critical for access to specific foods—particularly those that defend themselves by toughening or hardening their tissues. But they also show us that the types of defenses teeth have to overcome can be more important than the proportion of time spent on any given one. Extra-sharp or extra-strong teeth may be needed every day, as with Virunga mountain gorillas and Taï mangabeys. Or they may be needed only during especially tough or hard times that come with the changing seasons, as with Bai Hokou gorillas, or even every few years, as with the Kibale mangabeys. Paleontologists must keep these things in mind when they make their day-in-the-life dioramas for natural history museums. Paleoanthropologists must understand them when they consider the factors that led to the evolution of human diet.

This may not sound like a very big deal, but it is. It marks a fundamental change in the way those of us who study the fossil record think about adaptation. We cannot simply assume that fossil species with similar teeth had the same diets. Think of the brown lemurs and ringtails at Antserananomby. Likewise, we cannot assume those with different teeth ate different foods on a daily basis. Think of the mangabeys and red-tailed guenons at Kibale most of the time. Nature is much more complicated than that; and the implication for inferring foods eaten by fossil primates is profound.

Even if we understand how teeth work and what they are designed to break, that doesn't mean we know how individuals use them. We can no longer look at a handful of fossil teeth and say with great confidence, "Here we have a species that ate mostly leaves." Primates seldom eat all that they are capable of eating on a daily, weekly, or even monthly basis. There are many other factors that affect food choice, from competition with neighboring species to shifting availabilities of food types across time and space. We have learned in this chapter that those availabilities can fluctuate by season or even between years, and if a primate can't find what it wants on the table at the biospheric buffet, it has to eat what there is or leave hungry.

With all this in mind, let's go look at some fossils.

CHAPTER 3

Out of the Garden

If you've seen Stanley Kubrick's *2001: A Space Odyssey*, you know the story. It opens on the desiccated plains of Africa, two million years ago. There's a parched, starved band of "man-apes," as Arthur C. Clarke calls them in the book version, teetering on the brink of extinction. One picks up a limb bone from a skeleton on the ground and begins to smash its skull. The scene cuts to a tapir falling. The band eats. Clarke describes it a bit differently in the book. It's a pointed stone and a warthog, but the effect is the same. The man-apes look down on their hapless victim, "the future of a world waiting upon their decision."[1] Then comes a dawning awareness: our heroes realize they need never be hungry again.

It's a gripping tale. Drought robs the man-apes of the comfort and security of their lush ancestral forest home, leaving them hungry and vulnerable in the barren and desolate open country that replaced it. They meet the challenge and conquer nature with nothing more than determination, wit, and crude tools made from the remains of those less fortunate animals whose bones litter the savanna. A changing world sets our ancestors on the road to humanity, and diet takes center stage in the transformation.

The open savanna can be an unrelenting place in the dry season—stiflingly hot and dusty. The grassy plains and rolling hills are broken only by rocky outcrops with a few precious shade trees. There is the occasional meandering stream or small river lined by denser vegetation, but danger lurks everywhere—lions, hyenas, leopards, cheetahs. Imagine our ancestors in that unforgiving, pitiless world, and how it must have challenged them. But they answered the call. So now you're reading this book while, somewhere, a chimpanzee 98% genetically identical to you is climbing a tree in search of bugs, and contemplating little more than its navel. Its ancestor never left the rainforest with all its bounty.

We've certainly come a long way in the past few million years. How did we get here? Why are we so different from other primates? The environment must have played a role in human origins. That's how evolution works. We find most of our nearest living apish relatives in closed-canopy forest, but early hominin fossils are usually found in what is today much less hospitable open grassland. Was it like that millions of years ago when our distant ancestors walked the Earth? Did they move out into the savannas? Did the savannas spread out to meet them, or their descendants? Either way, at some point in the past there was a change in predators, competitors, and, of course, foods. How did our hominin forebears react?

We learned in chapter 2 that the biospheric buffet is ever changing, and that primates must choose from the items on offer at a given place and moment in time. Normal life-cycle events of plants, like fruiting, flowering, and flushing, all affect food availability over the course of a year, and availability can change dramatically between years according to longer-term cycles like the El Niño–La Niña Southern Oscillation. Imagine the effects of environmental fluctuations over thousands of years, or millions. At such long time scales, leaving the buffet hungry is simply not an option. An animal can often make it through the occasional lean period, even if it doesn't have the right teeth or guts to take full advantage of the foods that are available at the time. But if that turns into year after year, or even month after month, it becomes another story entirely. And century after century can mean extinction, unless a species evolves the adaptations it needs to survive on whatever new resources the changing world brings.

That's where teeth come in. As we learned in chapter 2, nature selects tooth size, shape, and structure for whatever foods a species is adapted to eat. These can be preferred foods or not, so long as there's an advantage gained by fine-tuning dental form and function. So, if a change in the environment means a change in menu options for a species, we might expect to find evidence for this in the teeth of an evolving lineage—including our own. In other words, teeth should provide us with clues about food choices in a changing world, and the roles of environmental dynamics and diet in making us human. Fortunately for us, there are plenty of teeth in the human fossil record to consider.

CHILDREN OF AFRICA

Some of the most important fossils in the world used to be kept in a small vault room at the University of the Witwatersrand (Wits) medical school in Johannesburg. The former dean of the faculty and my own academic grandfather, Phillip Tobias, used to call the collection "an embarrassment of riches." The vault door opened into a modest laboratory, a room lined by shelves packed with old plaster busts, the faces of human ancestors found throughout Africa and Asia, and casts of their teeth and bones. The replicas were there so researchers could compare them with the precious original specimens housed in the collection. There was a large table in the center of the room, covered with lime-green felt. That's where guests worked. I remember the first time I visited. I couldn't wait to see, and touch, the actual remains of our distant ancestors, ones who lived millions of years ago. I had read so much about them that each specimen was like a living, breathing being to me.

The custodian of the vault brought out a wooden tray cradling half a dozen specimens, teeth and jaws of *Australopithecus*. He laid it down on that lime-green felt. I had studied for years and flown halfway around the world for this moment. I was expecting it to be humbling and awe-inspiring, coming face-to-face with those fossils. But it wasn't; it was anticlimactic. They were just teeth—cold, lifeless teeth. I guess I had expected them to exude some sort of fundamental essence, some wisdom that told me who they were, what their lives were like, and how they were related to me. But they didn't.

Then it hit me. Everything I had learned about *Australopithecus*, its lifestyle and relationships, had been teased from those teeth and bones by the generations of researchers who came before. The fossils themselves weren't humbling, and they weren't wise. The scientists who gave them meaning were. This chapter introduces some of the key players in paleoanthropology, both the fossils and the researchers who have found and made sense of them. We consider how each has contributed to the understanding of our evolution, and the role of diet and the environment in making us human.

The Skull from Taung

Our story begins with Raymond Dart. I met him when I was a junior in college. It was at the Ancestors Symposium at the American Museum of Natural History in New York. All the great paleoanthropologists of the day were there, and I was witness to a historic event—one that brought together for the first time some of the most important original specimens in the human fossil record. At one point, when the audience filed out from the auditorium for a break between sessions, I headed to the restroom just outside. Too much coffee. I remember walking up to a urinal and glancing to my left. It was him, the grand old man who had proved six decades before that humans evolved in Africa! Dart had single-handedly changed the prevailing world view of the science. I was awestruck, and it was all I could do to walk back out and wait to introduce myself. He was very kind. We sat together on the floor outside the auditorium for the next half hour, and I listened to him recount the story of his discovery of the Taung child[2]—a defining moment in the history of human origins research.

Dart had taken a post as professor of anatomy at the newly minted Wits Medical School. He needed skeletons and fossils to build his teaching program and encouraged his students to collect them for him from South Africa's expansive open country, or *veld* (Afrikaans for "field"), over school holidays. In the summer of 1924, a student brought him the fossilized skull of a baboon she had gotten from a family friend. That friend was director of a limestone-mining company, and the skull had been found in a quarry near the village of Taung, about 250 miles southwest of Johannesburg. Limestone is great for preserving fossils, and quarrymen were hard at work blasting it out there, and elsewhere in South Africa, to supply lime for a booming gold-mining industry. Lime is used to process metal ores.

Dart immediately recognized that the skull was a big deal—the first fossil primate, at least to his knowledge, ever found in sub-Saharan Africa.[3] Were there more? Fortunately, one of the old miners at the quarry had taken an interest in fossils and had been collecting and saving them. Long story short—two large wooden crates arrived at Dart's home just as he was dressing for the wedding of a friend. He described the scene to me as if it had just happened the day before, as we sat together on the floor outside the auditorium in New York, 60 years later. He couldn't wait. He broke open the crates and found a fossilized brain cast—too large to be a baboon. This was even more important than he had originally thought. There he was—the best man, not yet fully dressed for the ceremony—digging through a crate filled with dusty, lime-covered fossils in search of the rest of the skull. He found it. The face was embedded in limestone, but it was definitely there.

Dart spent the next couple of months separating the skull from the rock that covered it, flaking away the limestone with his wife's knitting needles. The face was remarkable. It was nearly complete; its jaws still clenched as they had been in death. Its first molars were just erupting through the gums and into the oral cavity when it died. If it were human, it would have been about six years of age.[4] The face and jaw had humanlike parts, and it lacked the telltale large canines of a living ape. But the brain cast was stunning. The inside of a skull preserves some of the surface detail of the brain that presses against it in life.

The Taung child's skull had filled with sediment after burial, and a solution of calcium carbonate and water seeped in, which, over hundreds or thousands of years, formed the rocklike natural endocast. You can see the bumps and grooves of the brain's original right-side surface—even the artery that supplied the protective membrane that surrounded it. The opposite side of the endocast is covered in calcite crystals that glisten like diamonds in the light. More important, not only was the brain larger than that of a young chimpanzee or gorilla, but it had surface features that Dart thought were more like ours. And its stem emerged from the base of the brain, not the back end. This suggested to Dart a head held upright when the child was on two legs rather than four.

Dart named the fossil *Australopithecus africanus*, the southern ape of Africa, and proclaimed the Taung child to be of "an extinct race of apes *intermediate between living anthropoids and man*" (italics his).[5] But not everyone was convinced. The age of the deposit at the quarry was uncertain, and the exact location where the skull was found was unknown. In fact, Dart hadn't yet even separated the upper and lower jaws when he first announced the find, so it was impossible to have a good look at the tooth crowns, let alone compare them with those of other hominins. Also, the skull was that of a young child. What would

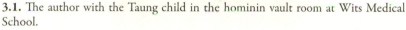

3.1. The author with the Taung child in the hominin vault room at Wits Medical School.

it have looked like as an adult? Finally, what about inferring it walked on two legs from the position of the brain stem? It seemed like a stretch to claim *Australopithecus* was a biped without any evidence from the neck down to support the argument.

In fact, the whole thing didn't make much sense at the time. Most researchers believed that humans evolved in Asia, or perhaps Europe, despite Darwin's speculation in *The Descent of Man* that "it is somewhat more probable that our early progenitors lived on the African continent than elsewhere."[6] The ape-man *"Pithecanthropus"* (now *Homo*) *erectus* had been found in Indonesia, there were Neandertals in Europe, and *"Eoanthropus dawsoni"* in England—all of which had much larger brains. *"Eoanthropus"* also had a much more primitive, apelike canine.[7] The Taung child's reported features just seemed wrong. If *Australopithecus* was really *"intermediate between living anthropoids and man,"* it should have more primitive canines than *Eoanthropus*, given its much smaller brain.

Sunrise on the Savanna

Dart wasn't making much headway persuading the scientific establishment by arguing from the basis of anatomy or geology. It was time
to bring out the big guns, the fundamental principles of evolution.
Charles Darwin had suggested half a century earlier that our ancestor's
descent from the trees owed to "a change in its manner of procuring
subsistence, or a change in the conditions of its native country."[8] But
the savanna hypothesis was all Dart. He was the first to reason that
human evolution required the challenges of a grassland environment—
the lack of water and food, the competitors and predators—for nature
to pressure our ancestors to evolve the deftness that comes with two
legs and the cunning that accompanies a larger, more humanlike brain.

But, back in the 1920s, it was thought that the climate of southern Africa hadn't really changed since the days of the dinosaurs, let
alone since the time of the Taung child. So if a dry and desolate habitat wouldn't come to our ancestors, they had to go to it. This was the
cornerstone of Dart's argument. Today's African apes, the chimpanzees
and gorillas, inhabit tropical forests to the north, separated from Taung
by the vast, dry Kalahari and open veld. No ordinary tree-hugging ape
could possibly have survived the trip across such forbidding and barren,
desolate landscapes. It would have taken a hardened, ground-dwelling
man-ape. The Taung child *had* to be a hominin, not just because of
where it lived, but because it was able to get there in the first place. Dart
wrote, "the formidable nature of the land and animal barrier together
with the vicissitudes of life . . . demanded the operation of choice and
cunning."[9]

Many remained skeptical that the Taung child, or the African veld
for that matter, had anything to do with human evolution. But Dart
did find an ally in Robert Broom, a Scottish-born physician and paleontologist who worked in South Africa's Karoo region at the time. Those
were the dry plains and bare hills to the south and west of Taung, well
known in paleontological circles for remarkable mammal-like reptile
fossils dated to between about 265 and 240 million years ago. Soon after Dart's announcement, Broom visited Johannesburg to see the Taung
child. He was convinced it was a hominin too, and within days had
submitted a short paper to *Nature* in support of Dart's conclusions.
This didn't help quell criticism much in the short term, but Broom

eventually silenced the critics. It would take a couple of decades and his own fossil finds to do it, though.

Broom moved to Pretoria in 1934 to become curator of physical anthropology and vertebrate paleontology at the Transvaal Museum.[10] There were lots of limestone quarries only an hour's drive away, around the Bloubank Valley just north of Krugersdorp, where Broom could search for his own man-ape. He set out soon after his move to "hunt for an adult specimen of the Taungs ape."[11]

Countless small abandoned lime quarries dot the landscape around what is today the Cradle of Humankind UNESCO World Heritage Site. The site straddles Gauteng and Northwest Provinces, only 15 miles west of the N1 and the sprawling corridor that connects Johannesburg and Pretoria. It includes 180 square miles of rolling hills covered in bushland and veld, with farms, game reserves, and the world's best chicken pies. There are also dozens of known fossil localities in the Cradle. These are cave deposits, now holes in the ground left by quarrymen after blasting for lime during the Witwatersrand Reef gold rush. The surface around them is today riddled with heaps of breccia blocks, piles of broken rock strewn about as refuse from the mining process. Each block is a hardened mass of soil, pebbles, and oftentimes bone, cemented together by calcium carbonate, which seeped down through the earth when the deposits were formed.

Broom visited one of those quarries, Sterkfontein, with a couple of Dart's students in 1936, and soon after found the first adult skull of an *Australopithecus*. The first bit of the rest of the skeleton, a thigh bone, was recovered there the following year.[12] And an entirely different type of hominin was found the year after that, at a site a couple of miles away, Kromdraai. Broom named the new hominin *Paranthropus robustus*, the robust near-human, because of its large and robustly built face, jaw, and cheek teeth. So, not only was there an adult version of the skull from Taung, but by mid-1938 there were other bones and at least two types of fossil hominin known from South Africa. The evidence could no longer be denied. We are indeed children of Africa.

Succulent Feasts for Murderous Beasts

As Dart continued to build his argument and piece together his evidence, the importance of an unforgiving but meat-filled savanna to human evolution began to take center stage. Dart believed that our ancestors' world changed when they left the primeval forest, and with that change they traded the life of a simple-minded, slothful frugivore for that of a cunning, industrious predator. His evidence was the broken fossil bones of animals found alongside those of *Australopithecus* in the quarries of South Africa. Dart thought they had the telltale fractures of bludgeoning, and he described in gory detail the techniques *Australopithecus* would have used to kill and eat their prey.[13] He wrote, "man's predecessors differed from living apes in being confirmed killers: carnivorous creatures that seized living quarries by violence, battered them to death, tore them limb from limb, slaking their ravenous thirst with the hot blood of victims and greedily devouring livid writhing flesh."[14] He was always the provocateur. When asked why he used such lurid prose in a scientific paper, he replied, "That will get 'em talking!"[15]

But his view of *Australopithecus* also came from analogy with the primates that live on the veld today. Dart reprinted a letter he had received from Harold Potter, then conservator at the Hluhluwe Game Reserve, in *The Predatory Transition from Ape to Man*. Monkey-rope Potter, as he was known, wrote of baboon troops hunting during the winter months, "possibly due to the fact that other food is then scarce."[16] If baboons on the ground in South Africa shift to hunting today when succulent plant foods are unavailable, *Australopithecus* certainly could have done the same in the past. Dart recognized the importance of seasonal change to primate diets decades before Melissa Remis, Joanna Lambert, or I was born, let alone had set foot in a forest (see chapter 2).

Actual evidence that seasonality may have played a role in the evolution of human diet, though, was still half a century away. We'll come back to that in chapter 5. So what *could* be said at that point about the diets of these early hominins beyond the presumed food remains found alongside them in the deposits and direct analogy with living primates? Robert Broom's protégé, John Robinson, was the first to consider in detail the hominin fossils themselves to work this out. He looked, unsurprisingly, to their teeth.

TEETH, DIET, AND A CHANGING WORLD

Robinson had not started out interested in teeth, or even human evolution for that matter.[17] He was working on plankton for his dissertation at the University of Cape Town in 1945 when he was offered a job at the Transvaal Museum to assist with curation of a newly acquired collection of moths.[18] While this wasn't quite what he had in mind for a career, World War II had just ended and jobs were scarce, so he accepted the position and moved to Pretoria. He soon after met Robert Broom and, three months later, joined him in the Department of Physical Anthropology and Vertebrate Paleontology. Broom was 80 years old at the time, and Robinson was only 23, but the two hit it off and began working together in earnest on fossils from the limestone quarries around the Bloubank Valley. They published nearly two dozen papers together before Broom's death five years later. These included the description of the jaw and teeth of a third type of fossil hominin, found by Robinson in 1949. It came from the site of Swartkrans, just across the valley from Sterkfontein. Swartkrans had already yielded *Paranthropus*, but this was clearly different. The teeth and jaw were smaller and more humanlike. Broom and Robinson named the new species "*Telanthropus capensis*," and proclaimed it to be intermediate between the other early hominins and today's humans. "*Telanthropus*" was later to be lumped together with "*Pithecanthropus*" from Asia in *Homo erectus*.

The stage was set and the three main players, *Australopithecus, Paranthropus*, and early *Homo*, were in place. How were they related to each other, and to us? What were their roles in human evolution? How did the evidence fit with Dart's savanna hypothesis? These questions all weighed heavily on Robinson's mind. He took over Broom's job after his elder colleague's death and focused his research in the early 1950s on the teeth of these hominins. Not only did he describe them in unprecedented detail, but Robinson also explained differences in tooth size, shape, and structure between the species by differences in diet. His was the first study to do this. Robinson was also the first to argue that two lineages of early hominins *could have* coexisted because of differences in what they ate and how they divvied up their habitats.

Compared with *Australopithecus, Paranthropus* had very large premolars and molars but relatively smaller front teeth. Also, its cheek teeth had very thick enamel and they wore flat, with scratches running side to side across the crown. Robinson figured that meant heavy use in grinding plant parts, like shoots and leaves, berries, and tough wild fruits. Its teeth were chipped too; perhaps, he thought, it ate grit-laden foods, like roots and bulbs. *Australopithecus*, on the other hand, had

A

B

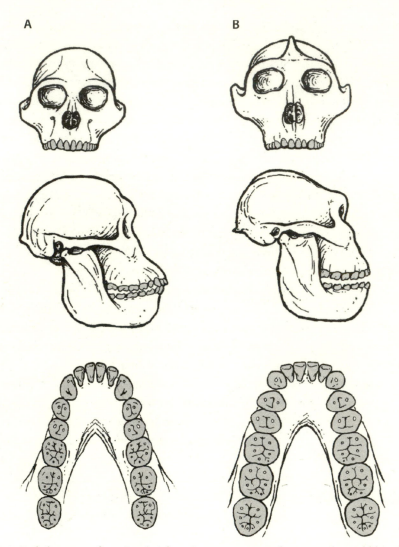

3.2. Early hominins from South Africa. Reconstructions of crania and mandibles of A. *Australopithecus africanus,* and B. *Paranthropus robustus.* Images courtesy of John Fleagle.

larger front teeth, including canines, and smaller premolars and molars. This seemed to him to match Dart's story better, and a varied diet that included meat. Robinson speculated that *Paranthropus* was the more primitive type, with *Australopithecus* having evolved toward the human condition. Early *Homo* had progressed even further, with yet larger front teeth and smaller molars. With that, John Robinson had laid the foundation for our current understanding of early hominin diets.

Robinson's next step was to try to understand what those differences in diet meant for the larger questions about human evolution, like the role of environment in the process. He needed context. He needed to know how the fossil sites formed, and what foods would have been around and available at the time. He needed details about the habitats in which the hominins lived. That job he gave to a young Rhodesian-born geology student, Charles Kimberlin (Bob) Brain. Brain began working on the geology of the hominin sites in the Bloubank Valley and another, called Makapansgat, about 150 miles to the northeast, in 1954. He earned his PhD for the study three years later, and soon after published it in "The Transvaal Ape-Man-Bearing Cave Deposits."[19] This work changed in a very fundamental way how paleoanthropologists of the day viewed the conditions under which our ancestors evolved. Had the veld, as Dart suggested, really remained unchanged since the dinosaurs roamed the Earth? Did the different hominins at the various sites live in the same habitats? These questions were key to interpreting differences between the teeth of *Australopithecus*, *Paranthropus*, and early *Homo*.

But where to start? Brain began by collecting soil samples in places around South Africa that differ today in the amount of rain they receive. Graskop, in the lush Drakensburg Mountains, averages 60 inches a year, whereas Alexander Bay, at the desert border with Namibia, gets less than 2 inches. The soils were different, and those differences made sense in terms of the rainfall and habitat each sample came from— heavily weathered sand grains for wetter conditions, lots of quartz dust for drier ones. He could use soil type, he reasoned, like a decoder ring to decipher rock samples from the hominin deposits. It looked as though the Sterkfontein and Makapansgat deposits had formed at drier times, whereas those at Swartkrans and Kromdraai accumulated during wetter ones. Since other extinct animals found at the first two sites seemed more primitive, Brain figured that Sterkfontein and Makapansgat were older than Swartkrans and Kromdraai. It looked as though conditions in the Transvaal had varied over time. The veld had not remained unchanging open grassland since the days of the dinosaurs after all. There was no single hominin habitat either. It seemed that *Australopithecus* had endured drier, veld-like conditions, whereas *Paranthropus* and early *Homo* lived during lusher periods.

This was just the kind of detail that Robinson needed. Geologists working in the Kalahari were also beginning to develop a broader view of past environments in southern Africa at the time. They believed that the Miocene epoch (see the geological time scale in figure 1.1) was mostly wet, with forests spreading across the region during much of the interval. But this all changed toward the end of the Miocene and into the Pliocene as the Kalahari began to form, with its red sands blanketing large parts of South Africa, Namibia, and Botswana.[20] The forests retreated, and conditions remained dry until the early Pleistocene, when rivers cut through the sands into the underlying limestone. The basic idea was wet to dry, then back to wet. The pieces of the paleoenvironmental puzzle were starting to come together. The main hominin deposits at Sterkfontein and Makapansgat, it seemed, fit into the middle, dry phase, and those at Swartkrans and Kromdraai formed during the wet one that followed.

Robinson began to build a story about the roles of teeth, diet, and a changing world in making us human. The sequence of events and general idea hadn't really changed much since Darwin proposed it nearly a century before, but now there were details. The first scene was about coming up on two legs, as the ancestors of *Australopithecus* left the forest—or, more likely as it seemed now, the forest left them. This would have freed their hands for carrying and using weapons and other tools, which, in turn, would have led to smaller front teeth. Canines and incisors would no longer be as important for fighting and feeding. But these early ancestors were still vegetarian, like living apes. They would have supplemented their diets with meat when they needed to, just as baboons do today, but they weren't driven from the trees by a bloodlust for flesh.

That came in the next scene. Shortly after Brain started working on the geology of the sites, he discovered some weathered pebbles at Makapansgat that looked as if they had been chipped into tools. He followed up with a careful search of the sites around the Bloubank River Valley, and found dozens of more convincing stone tools at Sterkfontein, several at Swartkrans, and a few at Kromdraai. Brain figured they were made by early *Homo*, because those at Sterkfontein (and the possible ones at Makapansgat) came from the upper layers, above the main *Australopithecus* deposits. It looked as though, by Swartkrans and

Kromdraai times, our ancestors had made the leap from bone-tool user to stone-tool maker.

Robinson considered the initial reliance on tools to have been triggered by longer dry seasons and more arid conditions. Meat, he thought, became important to our ancestors as they faced a drier, more barren homeland. More tool use and greater intelligence followed, as survival hinged increasingly on the successful hunt. This in turn would have led to a more humanlike lifestyle, which would have fed back into even more elaborate tool use, larger brains, and more efficient hunting. It would have been "virtually inevitable that adaptation would be carried well past the *Australopithecus* stage";[21] hence early *Homo*, with its increased brain size and the transition from a tool user to a toolmaker.

Paranthropus was different, of course. To Robinson, it also descended from the earliest hominin stock, but its extra-large cheek teeth and small front ones suggested it never advanced beyond the ancestral vegetarian diet. This fit well with wetter, lusher conditions in the Transvaal during Kromdraai and Swartkrans times. The fact that *Paranthropus* was found in the same levels as early *Homo* lent support to the idea that these hominins must have had very different lifestyles to coexist.

By the 1960s, the story was firmly entrenched: grasslands and deserts of southern Africa had fluctuated from wet to dry to wet again during the Plio-Pleistocene. *Australopithecus* evolved, eating progressively more meat, as drying conditions through the Pliocene led to a shift from a diet dominated by forest plants to one that included more animal protein. The other two hominins lived later, in the early Pleistocene, in the wetter conditions that followed. Early *Homo* carried the torch forward toward humanity, transitioning from tool user to toolmaker. *Paranthropus*, a late-surviving relic of the earliest hominins, was still largely vegetarian, despite its ancestors having climbed down from the trees and begun to use tools. Robinson's studies of tooth size, combined with Brain's work on paleohabitats, had brought diet and a changing world into the limelight.

The Present Informs the Past

The way researchers approached questions about human evolution, though, and the answers they got, were starting to change. The Second

World War had come and gone, and the global economy was growing. The new prosperity was a boon for science, a sort of "golden age," ushering in a new generation of young researchers with new ideas and new approaches. For paleoanthropology, it meant recognizing and documenting processes at work today, and using the knowledge gained to reconstruct the past—including relationships between teeth and diet, environment and evolution. Brain's studies on how rainfall affects soils in modern South Africa was one example.

Enter John Napier. Napier was a surgeon in London with a passion for anatomy and natural history—so much so that he started a primate research unit at the Royal Free Hospital Medical School. His approach was to use living primates as models to understand the hominin fossil record, just as Brain had used soils formed today to understand deposits at the Bloubank Valley sites. Napier's unit was a vibrant and exciting place, and many of the graduate students he trained went on to become academic superstars and drive the discipline forward.

Napier and one of his students, Colin Groves, revisited Robinson's idea that tooth size differences between the early hominins meant different diets. But they considered living apes, measured their tooth sizes, and looked for patterns to match anatomy with behavior. Chimpanzees and orangutans had large incisors and small molars for husking fruits and for pulping their flesh. Gorillas had smaller incisors and larger molars, for grinding "coarse vegetable matter."[22] The differences in tooth sizes between *Australopithecus* and early *Homo* were about the same as those between orangutans and gorillas. *Paranthropus* had even smaller front teeth relative to their back ones, suggesting consumption of even more tough vegetation. These lined up well with what Robinson had said and confirmed his general ideas—though Groves and Napier stopped short of explaining the differences in detail.

That was left to Cliff Jolly, another of Napier's students. Jolly had just completed his dissertation on fossil geladas and baboons from the same deposits as the early hominins. Baboons, and especially geladas, are today large, open-country monkeys. Many in the past were even larger, with extreme adaptations for life on the ground. Fossil geladas, like their living descendants, had a long thumb relative to their other fingers, smaller front teeth and large back ones, and heavy chewing muscles. Jolly's "ah-hah moment," as he calls it, came with the realization that these were some of the very same attributes that in early hominins had been associated with tool using and hunting. Geladas today use their hands and teeth for eating small, tough seeds. A large primate could do just fine on the ground without tools or hunting. Besides, Jane Goodall had just shown that forest-living chimpanzees, with very different teeth and hands, hunt and make tools at Gombe. A large primate living on the ground did not require tools to make a living, and tool use did not inevitably trigger hominin origins. There must have been something else. The answer for Jolly was environmental change.

Jolly published his seed-eater hypothesis in 1970. It built on Robinson's idea that the small front teeth and large molars of *Paranthropus* were those of a tough-food vegetarian. But it challenged his suggestion that hominin-like hands evolved for tool use. A gelada-like diet of small, tough seeds would do it. Jolly thought that the shift from a fruit-based diet to cereal grains could have begun in wetlands, and continued as hominins spread into wider floodplains and grasslands. Maybe this gave the hominins the anatomy they needed for meat eating and tool use. Even a slight environmental change might have been enough to prompt it. He mused, "perhaps an intensification of seasonality in a marginal tropical area which would put a premium on exploiting meat as an additional staple instead of an occasional treat."[23] Jolly believed at the time, as Robinson and Dart did before him, that *Australopithecus* had made that leap to meat eating given tools of animal bone, teeth, and horns; an enlarged brain; and improved bipedalism.[24] Early *Homo*, he argued, was further along the path.

The seed-eater hypothesis was built on conventional wisdom of the day—the fossil animals at the sites were food refuse of the early hominins, *Paranthropus* was primitive compared with *Australopithecus*, and Plio-Pleistocene habitats dithered with pendulous swings from wet to dry to wet to dry again. But by the time Jolly had developed his model, those ideas were already beginning to flounder. A wave of new finds, new technologies, and new ways of thinking about the past had begun with the discovery of hominins in eastern Africa. While South Africa had become increasingly isolated by apartheid since the days Broom and Robinson had worked the Bloubank Valley sites, fossils to the north brought more and more researchers from around the world to join the hunt. The way paleoanthropologists did business was beginning to change, and along with that came new insights into how a changing world made us human.

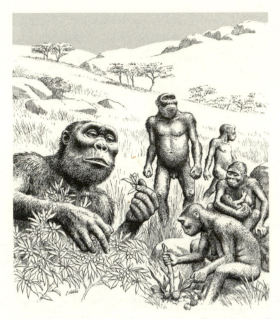

3.3. Reconstruction of *Paranthropus robustus* at Swartkrans in South Africa. Image courtesy of Fred Grine.

Eastward Bound

It started with Mary Leakey's discovery of a *Paranthropus* skull at Olduvai Gorge, where the Great Rift Valley passes through Tanzania, back in 1959. Mary's husband, Louis, named it *Zinjanthropus* at first, "Zinj" being the place name used by medieval Arab geographers for East Africa. To him this underscored the fact that South Africa was no longer the sole provenance of our forebears. Early *Homo* was found in the same deposits a couple of years later. That was huge because Olduvai was more than 1500 miles northeast of Makapansgat, and eastern Africa was now on the table. How many new fossil fields would be discovered? A paleontological "gold rush" followed and, by the mid-1970s, several more sites were found, stretching northward another 1000 miles (from Tanzania to Kenya to Ethiopia). Eastern Africa had produced all three types of hominin—*Australopithecus*, *Paranthropus*, and early *Homo*—albeit new species. The sequence of their appearance in the fossil record was the same too: first *Australopithecus*, and then *Paranthropus* and early *Homo*. And just as in South Africa, tools were found only in the Pleistocene-age deposits that produced the latter two.

But there was an important difference between the eastern African and South African hominins—new methods for dating volcanic ash allowed geologists to determine the actual ages of hominin-bearing deposits in eastern Africa. Before then, no one *really* knew how long ago the hominins, or other fossil species for that matter, lived. Geologists in Raymond Dart's and Robert Broom's day could figure out a rough sequence within a site, and they could argue that deposits at different sites were of similar age if they contained similar fossils, but everything else was guesswork. How long did it take for the layers to pile up? Most paleontologists at the time guessed that the Pleistocene epoch started around 500 thousand to 600 thousand years ago (kya). But actual dates would put our antiquity in its proper place. They would give us a real sense of our time on this planet. They would allow us to figure out not just when *Australopithecus*, *Paranthropus*, and early *Homo* lived and how much time there was between species, but also the timing of changes in anatomy related to function, and how they lined up with the geological evidence for climate.

Berkeley geologist Jack Evernden first visited the Leakey camp at Olduvai Gorge to collect samples for dating in 1957. Eastern Africa had plenty of active volcanoes during the Plio-Pleistocene, and there were lots of tuffs, rocks made from volcanic ash, at Olduvai to date. In fact, the site and others in eastern Africa are nothing like the limestone quarries to the south. The Olduvai deposits formed as blankets of soft sediment and bone laid, one on top of the other, over a hard, volcanic bedrock. The gorge has four main fossil-bearing strata, divided by sheets of ash, which form like the icing that separates layers of cake. Two main ravines, each about 300 feet deep, cut down through those layers, exposing the deposits over a stretch of several miles. We'll learn more about Olduvai later.

Evernden took his first tuff samples near the bottom of the gorge. They gave him a date of 1.75 million years ago (mya). But no hominin would be found at Olduvai for another couple of years, so they didn't get all that much attention. Once *Paranthropus* was discovered there, though, Evernden's colleague Garniss Curtis visited the site to collect more samples. He confirmed the date, 1.75 mya. This was a game-changing discovery. *Paranthropus* was much older than most had imagined. The best guess beforehand was that the Olduvai hominins lived about half a million years ago. In one fell swoop, the span of human evolution was nearly quadrupled! With that, the course of human evolution, the sequence from one species to the next and environmental changes that triggered it all, started to fall into place.

The discovery of *Paranthropus* at Olduvai also led to a new view of relationships between *Australopithecus*, *Paranthropus*, and early *Homo*. Louis Leakey asked Phillip Tobias, Raymond Dart's protégé and newly appointed successor at Wits, to describe and analyze the skull his wife, Mary, had found, soon after its discovery in 1959. After years of careful analysis and comparisons with the South African hominins, Tobias produced a huge monograph on the specimen. His conclusion was that *Paranthropus* was not the last gasp of a phantom lineage more primitive than *Australopithecus*, as Robinson had envisioned. Its small canines, giant cheek teeth, and related features of the chewing system were actually highly specialized. On the contrary, a reanalysis of *Australopithecus*'s brain size, along with its smaller molars and larger canine teeth, suggested that the earlier hominin was actually more primitive. In fact, Tobias believed that the ancestor of early *Homo* and *Paranthropus* was "virtually indistinguishable" from *Australopithecus africanus*.[25]

At about the same time, more and more hominin sites were being discovered in eastern Africa. A massive joint French, Kenyan, and American research expedition to the lower part of the Omo River Valley in southern Ethiopia between the late 1960s and mid-1970s produced something like 50,000 fossils, including more than 200 hominin specimens. Most came from a formation called the Shungura, and geologists worked hard to piece together the ages of its deposits. In the end, they came up with 25 distinctly dated layers ranging from about 3.6 to 1.05 mya. Dating fossil deposits in the Omo was an enormous undertaking, but it ultimately allowed researchers to date not only the fossil hominins found there, but also those in South Africa and elsewhere.

Looking Back South

Meanwhile, back in South Africa, there were no volcanic tuffs to date. There were plenty of fossil animals though, and paleontologists had a general idea of the order in which the various species evolved. Researchers in the 1950s had worked hard to match deposits at the various sites, both those within South Africa and those in other parts of the world, based on their fossil species. The hominins all seemed to be early Pleistocene, with the bulk of those at Sterkfontein and Makapansgat a bit older than the ones at Swartkrans and Kromdraai. And soon after *Paranthropus* was found at Olduvai, researchers started to compare the other fossil animals there to those at Swartkrans and Kromdraai. They matched pretty well. Now researchers were starting to get somewhere! They could use dates from the sites in eastern Africa to get an idea of the ages of early hominins in South Africa by looking for common fossil animal species.

Basil Cooke, a South African–born geologist then at Dalhousie University in Halifax, Nova Scotia, took on the task. Cooke had worked at Sterkfontein three decades earlier, while he was on the faculty at Wits, and knew the hominin deposits well. He began with fossil pigs from the Omo expedition to match species with those marvelously dated layers. Cooke found that, over time, pig third molars got longer. In fact, the relationship between tooth length and date was so good that he could reasonably estimate the age of the South African hominin deposits by measuring the molars of fossil pigs found in them. Elephant back teeth could be used too. Vincent Maglio at Princeton had found their crowns got taller and developed more ridges over evolutionary time. The pigs and elephants together suggested the main hominin-bearing deposits at Sterkfontein and Makapansgat actually dated to the late Pliocene, about 2.5–3 mya, whereas those at Swartkrans and Kromdraai were indeed early Pleistocene, dating to around 2 mya, perhaps a little younger.

By the mid-1970s, *Australopithecus* (the Lucy species—*Au. afarensis*[26]) had been found at Laetoli, 20 miles from Olduvai, and also 1000 miles north from there, in the Afar Triangle of Ethiopia. The Laetoli hominins dated to 3.6–3.8 mya, and those in the Afar were a bit younger, albeit still older than about 3 mya. So not only did Phillip Tobias's study of the fossils suggest *Australopithecus* could have evolved into *Paranthropus* and *Homo*, there was also plenty of time for it to have happened.

New ideas and approaches to interpreting the fossils also began to surface as more young scientists were drawn into the discipline. Some started to question Dart's interpretation of how the animal bones and teeth found at the sites had gotten there in the first place. Were they really the food refuse of early hominins, or might they *and the hominins* have been dropped by carnivores or scavengers, like big cats or hyenas? In other words, was *Australopithecus* really the hunter, or was it the hunted? Brain had left South Africa for a position as deputy director of the National Museum in his native Rhodesia (now Zimbabwe) in 1961, following a falling out with Robinson. But these questions drew him back. He returned to fill his old mentor's post at the Transvaal Museum four years later, after Robinson left for the University of Wisconsin.

Brain's approach to addressing the "hunters or hunted" question laid the foundation for an entire discipline, *taphonomy*, which is the study of what happens between the moment an animal dies and when we dig it up, or blast it out of a cave. His answers systematically disemboweled Dart's interpretation of how the fossils in the South African sites had gotten there, the view of *Australopithecus* as a predator, and the role of hunting in early human evolution. He started by reconstructing the environment around Swartkrans when the fossils were deposited. The bones at the site had accumulated in an underground cavern connected to the surface by a vertical shaft about 5 feet across and 50 feet long. Hominins would not likely have lived or eaten there, deep in the dark zone. The bones probably fell in, or were washed in following rains. Judging from the telltale gnaw marks, breakage patterns, and types of bone found, both those of hominins and other animals, these were in fact mostly the unfortunate victims of leopards, hyenas, or saber-toothed cats. The hominins were, like the other prey animals at the sites, eaten by carnivores. Hominins may have contributed to the accumulations, especially given stone tools found in the more recent layers, but the Plio-Pleistocene cave deposits were certainly not the butcher shops or dining halls of *Australopithecus* and early *Homo*.

HOME ON THE RANGE

By the mid-1970s, researchers were starting to question many of the long-held, fundamental beliefs of those who came before them. Did the Bloubank Valley *really* get wetter and lusher between Sterkfontein and Swartkrans times? That didn't make sense if Jolly's seed-eater hypothesis was right. Gelada monkeys today live in dry, mostly treeless mountain meadows. Would *Paranthropus* really have evolved a gelada-like diet just at the time forests were beginning to fill in and close the landscape? Would early *Homo* emerge, and become a big-game hunter, if grasslands were actually becoming more lushly vegetated? Elisabeth Vrba, a high-school teacher turned fossil antelope expert, didn't think so.

Vrba had majored in zoology and mathematical statistics in Cape Town in the early 1960s and, like John Robinson two decades earlier, had hoped to become a marine biologist. Finances got in the way, though, and she left Cape Town for Pretoria to teach high school.[27] She became interested in paleontology, and within a year had visited Bob Brain at the Transvaal Museum in search of opportunities to work with fossils. Brain gave Vrba her start, at first as an unpaid assistant, preparing and sorting the enormous collection of fossil bovids, antelope and their kin, that had piled up from the Bloubank Valley sites. Freeing bones and teeth from the breccia that encased them was a time-consuming and sometimes insipid task, but it set her on the path to develop enormously innovative and influential ideas about how the environment drove evolution.

Vrba couldn't have chosen a better family of mammals to study. First, bovids are easy to sort. We can recognize an antelope by its horns; look in any safari guide. The bony shafts that anchor those horns to the skull are also distinctive and allow paleontologists to identify and separate species. And there were a lot of them. Bovids are often the most common large mammals at early hominin sites, both those in the Bloubank Valley and those in other parts of Africa. So, like Cooke's pigs and Maglio's elephants, they could be used to further refine dates for the South African hominins. But unlike pigs and elephants, there are about 75 species of bovid living in Africa today. The important thing about them for our story, though, is that most are particular about where they live. Some eat grass in more open settings, and others eat tree and bush parts in more wooded or forested habitats. Vrba could use fossil bovids found in the same deposits as early hominins to get a sense of the environments in which they all lived.

She started by assuming that bovid species in the past ate the same sorts of foods that their descendants do today, and that they lived in the same sorts of places. Most of the dozens of species of wildebeest, gazelle, and their near kin are found grazing in open grasslands, for example. And Vrba found that the proportion of these bovids relative to other types in an area today is tightly bound with the ratio of grass to trees. Because grass-eating gazelles and wildebeests dominate bovid communities on African savannas today, the idea was that their proportion at fossil sites could serve as a proxy for habitat in the past. This was to become the basis of Vrba's doctoral dissertation in 1974, as well as much of her work since.

It was time for the first test. If Brain's studies of fossil soils were right, *Australopithecus* lived in drier conditions than *Paranthropus* and early *Homo*, and there should be more gazelles and wildebeests in the earlier deposits at Sterkfontein and Makapansgat than at Swartkrans or Kromdraai. But there weren't. In fact, Vrba found the *opposite*. Antelopes in *Australopithecus* deposits were actually more adapted to forest and bush, and those found with *Paranthropus* and early *Homo* remains were adapted to grassland. Brain, it seemed, had correctly identified climate change, but he had apparently gotten the details backward! But how? He had assumed that sediments in the caves reflected the climate at the time they were laid down. Soil formed, then winds kicked it up and carried resulting dust into the caves. In reality, though, wind-borne dust is more common with erosion and exposure of old surfaces, those formed long beforehand. In fact, erosion had cut deep into the Kalahari sands, exposing the earlier, wetter African surface in places. The winds had actually carried sediments from those more ancient times to Swartkrans and Kromdraai. That's why they looked wetter.

Vrba's South African bovids suggested instead a drying trend over time, rather than the climatic dithering Brain had inferred from the rocks. Increasingly arid conditions also seemed to have developed in eastern Africa during the Plio-Pleistocene. Remember the pattern of longer molars in pigs, and higher-crowned ones in elephants, over time? That fits better with the idea of a progressively more open landscape and more grit-laden, abrasive grass. It also makes more sense in terms of how researchers at the time were interpreting changes in early hominin diet, both the gelada-like specialization for *Paranthropus* and increasing reliance on hunting by early *Homo*. Vrba's work had tied environment, diet, and hominin evolution together with a nice, neat bow. The spreading savanna, or veld, at the onset of the Pleistocene seemed to drive the origin and evolution of these two hominin lineages. You could see it in the changes in diet over time.

Vishnu, Siva, and Brahma

Vrba continued to study fossil bovids over the next decade as she moved up through the ranks at the Transvaal Museum. She found that they had more to teach us about human evolution. Antelopes are, as early hominins were, large-bodied, plant-eating mammals that live on the savanna and in the forests of Africa. But they were a lot more common than hominins, and there were many more species of them. Much could be learned from bovids, and applied to hominins, about the way evolution worked.

But researchers up to that point had put little effort into using fossils to understand the process. Most thought the fossil record was too incomplete to get at the details. That changed in the 1970s, when Niles Eldridge at the American Museum of Natural History and Stephen Jay Gould at Harvard argued that the record was, in some cases, just fine for working out the tempo of evolution. It looked to them as though many species appeared suddenly, persisted for some time unchanged, and then disappeared as quickly as they had come. Was this pattern real, or simply an artifact of an imperfect fossil record? In other words, did evolution occur in fits and spurts, by what they called *punctuated equilibrium*, or did it proceed gradually and continuously as competing species jockeyed for resources and struggled to survive, as Darwin had originally proposed?[28] Eldridge and Gould's ideas were huge at the time, and everyone was talking about them.

I was taught that species change when new, beneficial genes are introduced through mutation, and individuals with them out-reproduce those without. This is Darwin's *survival of the fittest*. But what are the chances a mutation in one or even a few individuals would spread to thousands or millions of descendants? Eldridge and Gould reasoned that even a helpful trait would likely be absorbed in the population, swamped by other gene variants. So just by virtue of the number of individuals in a species, evolution should be rare. And when it does happen, it should occur at the outer edge of a species distribution, where there is less mixing of genes, and there are smaller populations and more marginal environments. Edge populations could more easily become isolated from the mainstream and bud off to form new species. If evolution was happening in such small, marginal populations, as Eldridge and Gould reasoned, it's no wonder that we often don't catch the intermediate forms. It's no wonder that the fossil record is full of species that seem to appear, persist for a time unchanged, and disappear.

But while Eldridge and Gould put a lot of effort into working out *how* evolution happened, they didn't spend much time trying to figure out *why*. What drives new species to come and go? That was the next big question Vrba was to tackle. She reasoned that time alone wouldn't do it; there must be some sort of trigger if Eldridge and Gould were right. All those years working on bovid bones and teeth made it clear to Vrba that environmental change was that trigger. She wrote in 1980, environment is "the 'motor' of evolutionary change."[29] The first clue was that there are fewer generalist species, those that are equally happy with grass as with trees and bushes, than there are specialists. In terms of number of species, the specialists were more successful. But were they *really*? That depends on how you look at it.

Generalists can survive and flourish in just about any setting. But specialists tend to be much less comfortable with habitat change. Take the spread of grasslands across southern and eastern Africa, for example. As the region dried out, savannas would have spread to create a patchwork of wooded islands separated from one another by a sea of grass. Generalists could weather the storm of climate change, thrive, and even spread, to eat whatever nature's bounty had to offer wherever they happened to be. The pickier specialist species, on the other hand, would have fragmented into smaller populations, separated and isolated in patches containing the food they were willing and able to eat. I think of the large scientific conferences I attend. There are thousands of us in one place. Some seem not to care what they eat, and are driven to restaurants farther from the conference center only by the length of the wait for a table. These folks distribute themselves in decreasing numbers with distance from the bar at the meetings hotel, where everyone seems to congregate at the end of the day. Others are fussier and have specific preferences. Their distribution is much patchier and depends on the type of food they want to eat—Ethiopian, Indian, barbeque, or whatever.

When environmental change results in a patchy environment, specialists become isolated into small populations ripe for evolution. Generalists can keep themselves spread out. Vrba reasoned that this was why there are dozens of specialist bovids in Africa today, like gazelles and wildebeests, but just a few generalists. She had discovered the "why" behind Eldridge and Gould's model of how evolution worked. Vrba explained it using a Hindu triad analogy. There was Vishnu, the preserver. If there are no barriers to migration, species can move to find the food they need without changing. Then there was Siva, the destroyer. Fragmented populations can shrink themselves into extinction. Finally, there was Brahma, the creator. Species can evolve to meet the challenges of their new settings. In other words, species that can't handle an environmental makeover have three options: move, die, or change.

Pulses of Turnover

Vrba also realized that if a change in the environment is substantial enough to cause migration, extinction, or evolution in bovids, it should trigger the same reaction in other lineages at the same time. There should be synchronized bursts, or pulses. Perhaps this explained the turnover of both bovid and hominin species with the onset of the Pleistocene epoch between Sterkfontein and Swartkrans times. Not only are there more dry-adapted bovids, but *Paranthropus* and *Homo* appear too. Vrba had tied together Eldridge and Gould's punctuated equilibrium theory, which was the rage at the time, with human evolution, by suggesting that environmental change caused a community-wide turnover of species. She announced the turnover-pulse hypothesis at a conference in Nice in 1982, and people were still buzzing about it two years later at the Ancestors Symposium I attended in New York. Not only was it an intuitive and logical idea, but it was testable. So the search was on for turnover in other lineages and for the environmental triggers that caused them.

The search was also on for other turnover events. Vrba reasoned that if an environmental change triggered the origins of *Homo* and *Paranthropus*, it must have caused other events in human evolution at other times too. Several new bovids evolved at the Miocene-Pliocene boundary, around the same time the human and chimpanzee lineages were thought to have split. There seemed to be another pulse of antelope evolution just under a million years ago—perhaps around the time that *Paranthropus* went extinct and *H. erectus* was (then) thought to have spread through Asia.[30]

FINAL THOUGHTS

The idea that the environment played an important role in our evolution is not a new one. It started with Dart's belief that crossing the treacherous open plains between central African forest and South African veld gave early hominins the challenges and opportunities that set them on the course to humanity. Recall how Robinson looked to fossil hominin teeth to understand dietary differences between the species, and how Brain tied those differences to environmental change over time with habitat reconstructions. We shouldn't forget Jolly's work on living geladas as analogs for the early hominins, either. That made clear the importance of the environment as a trigger for human evolution. Finally, as new fossils were unearthed in South Africa and up the Great Rift Valley, and precise ways of dating the sediments that entombed them were developed, the role of environmental change in human evolution came into better focus, culminating in Vrba's turnover-pulse hypothesis.

Vrba's work inspired a whole new direction for human origins studies—the search for evidence to tie specific events in our evolution to large-scale changes in climate. Researchers looked to the appearance of the West Antarctic Ice Sheet, the onset of the Pleistocene ice ages in the northern hemisphere, and so on. And it just so happens that an important new field of research, paleoclimatology, was coming of age at about the same time. A perfect storm was brewing, and things were about to get interesting.

CHAPTER 4

Our Changing World

The Sahara stretches 3000 miles across North Africa, from the Atlantic Ocean to the Red Sea. Its landscapes are among the hottest and driest on the planet. Temperatures soar to more than 130°F; rainfall averages less than three inches a year; and intense, almost suffocating, wind gusts sandblast everything in their wake. The plants and animals that call the great desert home must be hearty, drought-resistant species. While the Sahara is nearly the size of the United States, it supports less than one tenth the human population. But it wasn't always a harsh and forbidding place. You can tell by the rock art. Buffalo, rhino, antelope, elephants, hippos, and many other animals are painted on or carved into thousands of cave walls and rocky outcrops. The Sahara 5000 years ago was teeming with animals, lush vegetation, and overflowing lakes. And it will be green again. Some say the change back has already begun.

This is a wake-up call for anyone trying to understand how a changing world made us human. Massive fluctuations in habitat can occur in what amounts to little more than a tick of the clock when your time scale is measured in millions of years. Those fluctuations may be extreme in some places, like the Sahara, but hardly noticeable in others. Yes, we evolved on a changing landscape, and, yes, environmental change triggers evolution. While the big picture is reasonably straightforward, the details are often not.

It all seemed so much simpler back when I was in graduate school. A series of cooling and drying pulses spread across the planet over the past several million years. One after the other they struck, like waves hitting a shoreline as the tide rises. Each swept away more of Africa's primeval forest and its hapless inhabitants. Each left savanna in its wake to be inherited by those that survived, those that were transformed in the process, including our hominin ancestors. It was an elegant hypothesis. Our planet has, on the whole, trended cooler and drier with a series of steplike changes since the Miocene. Each step marked an important milestone on the road to humanity—our split with the chimpanzees, the earliest members of our genus, the spread of *Homo* out of Africa. Researchers looked to the advance of glaciers, to the spread of cold steppe grasslands across Europe and Asia, and to growth of the Antarctic ice sheet for evidence. They documented global-level changes and long-term trends in climate.

A big picture was emerging, but so too was concern about its resolution. Was there enough detail to tie together environmental change and human evolution? The Sahara example makes us wonder. So does global warming today. While the average temperature of the planet is up and many places are getting warmer, some spots are actually getting cooler. Some are wetter, and others are drier. Many places are developing more variable or extreme weather. Yes, there are global-level patterns, but the local effects are complicated. They become even more so over the time scale of human evolution because the Earth's surface constantly reshapes itself—erosion and weathering, deposition of sediments, and volcanic eruptions. Mountains and valleys form on land and in the sea as the plates that underlie continents and oceans shift. These can change how air and water circulate, and can have dramatic effects on local conditions.

Fortunately, paleoclimatologists have developed some amazing new tools to document past fluctuations in global climate and changes in regional and local environments. It has required painstaking work by teams of researchers around the world, but the details are beginning to come into focus. And they are transforming the way we think about human evolution.

This chapter introduces a few of the key players and chronicles the extraordinary lengths to which they go to build a record, an archive as paleoclimatologists call it, of climate and environmental changes through Earth's history. Some measure those changes in millions of years and others use millennia, centuries, or decades. Some consider continent-wide patterns and others concentrate on individual lake basins, river valleys, or fossil sites. These different scales of time and space give us different pictures, but all contribute to our understanding of the habitats of the early hominins, the foods available to them, and how those affected the course of our evolution.

The more we learn, the more complex the Earth system seems to be.[1] Global average temperature and precipitation levels swing back and forth, thanks to an elaborate dance between the Earth and Sun. The way climate affects the environment varies with local conditions, which themselves change as tectonic plates shift and the Earth reshapes itself. When we put all this together with the fossil record, it's impossible to escape the conclusion that our changing world made us human. But how? Our paleoclimate archives are only now beginning to approach the resolution we need to begin to address that question. It's already obvious, however, that relationships between climate change and human evolution are more complicated than originally envisaged. Let's take a look.

THE DANCE BETWEEN EARTH AND SUN

"Nothing endures but change." That statement is just as true today as it was 2500 years ago when Heraclitus of Ephesus said it. The Earth's climates have been changing ever since our planet first formed four and a half billion years ago, and they will still be changing when the expanding Sun engulfs it in seven and a half billion more. It's happening all around us. Average sea and land surface temperatures are on the rise and the ice caps are melting. Extreme weather is becoming more severe, and heavy precipitation events are more common, even in places where total rainfall is down.[2] Few scientists doubt that this is our fault. The carbon dioxide and other pollutants we're pumping into the atmosphere are blanketing our planet and trapping the Sun's heat. But what about natural climate change? How does that work? Much of it comes down to orbital dynamics—cyclical changes in the tilt of the Earth and its distance from the Sun. Africa's monsoon climate system gives us a great example of how the relationship between Earth and Sun affects our world.

The African Monsoons

The African monsoons are not simply torrential downpours in exotic rainforests and deserts. They are cycles of seasonal weather change, brought on by the simple fact that land heats and cools more quickly than water. When hot air rises, it creates an area of low pressure. When cool air falls, high pressure follows. It makes sense, then, that there would be lower air pressure over sub-Saharan Africa during warmer months than over the adjacent eastern Atlantic and western Indian Oceans. Because currents flow from high pressure to low, moisture-rich sea breezes blow inland, dropping rain as they rise. The pattern is reversed at cooler times of the year, when drier conditions on the continent thin vegetation cover and expose soils to wind erosion. Continental breezes follow the low air pressure offshore, blowing half a billion tons of dust from Africa into the seas each year.[3]

In some places the effect is dramatic: drenching rains and lashing dust storms. During the rainy season, ephemeral lakes fill and rivers flow. The grass turns green, and the savanna comes to life. It becomes a world of plenty. But in the dry season, that same world can be a lifeless, desiccated plain, as those who are able to leave follow the rains to lusher places. The height of the Sun in the sky and variations in how quickly land and water heat and cool can be the difference between life and death.

In other places, the effect is subtler, though no less important to those who live there. Recall from chapter 2 the seasonal shortage of fruit for the gorillas at Bai Hokou, and the ever-changing options on the biospheric buffet for apes and monkeys at Ketambe. Both of these sites are just north of the equator and have dry seasons from December to February, when the Sun's rays are most diffuse. At Bai Hokou and Ketambe the connections between weather pattern, plant life-cycle events, food availability, and diet are clear. As we've learned, feeding ecology and the role of a primate in the larger community of life depend on both the accessibility that teeth and guts afford and what's available to eat at a given place and time.

The seasonal cycle of shifting winds and the rain or dust they bring is driven by the Sun. When it is overhead, the temperature imbalance favors rain over land, and when it's not, drier conditions prevail. This relates to the Earth's tilt. Think of a flashlight pointed toward a wall at night. If the beam strikes head on, the light is concentrated in a small circle. But if it skims the wall at an angle, it becomes drawn out and diffuses as it spreads over a larger surface. That means less energy striking any given area on the wall. Imagine that same beam striking a curved surface, say a basketball, from a distance. The light would be most concentrated on the point closest to the source, and more diffuse as it spreads outward.

The Earth works the same way. Because the equator gets more direct sunlight than the poles, more of the Sun's energy ends up being absorbed there. This leads to an imbalance, intensified by the fact that snow and ice reflect more solar radiation back into space than do darker oceans and tropical vegetation. We can think of climate as the Earth's reaction; air and sea currents redistribute heat around the globe in an attempt to balance the system.[4] But where do the seasons come in? The Earth revolves around the Sun with its axis fixed at an angle of 23.5 degrees from the plane of its orbit, pointed in the north nearly directly at the star Polaris. Because that doesn't change much as the Earth moves in orbit, the tilt of the north pole shifts from toward the Sun, to away, then back toward it over the course of a year. During the summer solstice, the northern hemisphere gets more direct sunlight, with warmer temperatures over longer days. Half an orbit later, the northern hemisphere is tilted away, resulting in more diffuse solar radiation over the area, with shorter days and cooler conditions. A steeper angle of approach for the winter Sun's rays also means more atmosphere to get through, which in turn means more solar energy reflected back out into space before it can hit the surface.

Milankovitch Cycles

If atmospheric and ocean currents and the changing seasons are driven by the dance between Earth and Sun, surely changes to our planet's orbit and distance from its star must have a dramatic effect on climate. The patron saint of paleoclimatology, Milutin Milanković, worked out the details early in the twentieth century. Milanković had begun his career in civil engineering, designing concrete dams, bridges, and viaducts. These earned him fame, fortune, and a chair in applied mathematics at the University of Belgrade in 1909. He began working on orbital dynamics and their effects on climate in 1912, and published his seminal work, *Mathematical Climatology and the Astronomical Theory of Climate Change*, in 1930.[5]

Milanković identified three important aspects of movements between our planet and its star that affect climate: *eccentricity, obliquity,* and *precession*. Eccentricity is the shape of the Earth's orbit, which varies from nearly circular to elliptical, in cycles of about 100 and 400 millennia. At maximum eccentricity, the distance between Earth and Sun fluctuates by more than 10 million miles, and the annual variation in solar radiation hitting a given point on the globe is greatest. Obliquity is the tilt of the Earth's axis, which varies by nearly three degrees over a 41,000-year cycle. The greater the tilt, the more distinct the seasons. Finally, precession is the wobble of the Earth's rotational axis. The north pole gyrates, like a top losing its momentum, over a 27,000-year cycle. We call these cycles of eccentricity, obliquity, and precession *Milankovitch cycles*. When we combine them, we get the closest approach of the Earth to the Sun coinciding with maximum tilt toward the Sun every 19,000 to 23,000 years.

Orbital dynamics are of interest to us here because they throw the Earth's climate system out of whack. Let's take another look at the Sahara. The June solstice currently occurs just a few days before the Earth is farthest from the Sun. That means less solar energy to heat the Sahara during the northern hemisphere summer, which means less of a gradient in atmospheric pressure between land and ocean to draw in rains. But 10 million years ago, the Earth was closer to the Sun in June, and summer monsoonal rains soaked the landscape. The transformation of the Sahara five million years ago, then, was a natural outcome of orbital precession.[6]

Milanković himself focused on the northern hemisphere ice ages. He reasoned that cooler summers meant less ice melt, so that glaciers could persist throughout the year. Northern hemisphere ice ages should be a natural outcome of decreasing the tilt of the Earth to minimize seasonal extremes, and maximizing the distance between Earth and Sun during the summer months. That seems to be just what happened. But how did Milankovitch cycles affect the tropics? If we're interested in how climate change drove human evolution, we need to understand the effects of orbital dynamics on conditions in sub-Saharan Africa.

Dust in the Wind

This was very much on the mind of Peter deMenocal in the mid-1980s. He was a graduate student at Columbia University at the time, studying

4.1. *JOIDES Resolution.* Image courtesy of the Integrated Ocean Drilling Program.

magnetic properties of sediments from the floor of the Atlantic Ocean off the western coast of Africa. Because continental dust is easily magnetized, the magnetic properties of offshore sediments should provide a reasonable estimate of the intensity and length of dry seasons in subtropical Africa. In principle, variation in magnetic susceptibility along and between cores could be compared with climate reconstructions for higher latitudes, and matched up with changes in the Earth's orbit and tilt.

DeMenocal's big break came when a spot opened up on the *JOIDES Resolution*,[7] a scientific drilling ship that has been exploring the world's ocean floors for the past three decades. It was leg 117 of the ocean drilling project, an eight-week cruise set to recover ocean-bottom sediments from the northwest Arabian Sea off the east coast of Africa. A more senior scientist had been scheduled to go, but he had to bow out because his wife was expecting. "I suppose it was fate," deMenocal recalls. They needed a paleomagnetist, and he was able and ready, so deMenocal, along with 27 other scientists from around the world and a host of support staff, set sail from Sri Lanka in August of 1989.

The *Resolution* is 470 feet long, with a 200-foot-tall derrick towering over the middle of the ship. The drill team fits massive 100-foot stands of steel pipe together and lowers them to the sea floor, the resulting tube connecting ocean bottom to ship being miles in length and hundreds of tons in weight. It's a colossal effort. A piston is lowered inside the tube and held in position by three thin pins as sea water is pumped in on top. When the pressure builds, the pins break, and the weight of the water smashes the pipe down ever deeper into the ocean floor. Then each core is raised to the surface, 30 feet at a time, and cut into sections roughly 4.5 feet long. The process takes only 15 minutes per core, and the scientists struggle to keep up, processing core after core. Life aboard the *Resolution* has been described as 90% boredom, 10% action,[8] but to witness dust blown from the African continent seeing the light of day for the first time in millions of years is worth the wait for those fortunate enough to be there. In total, 22,000 samples were retrieved from 25 holes at a dozen sites during leg 117.

Even between drilling sessions, the geologists aboard felt like kids in a candy store. The *Resolution* is a floating research center, five decks equipped with state-of-the art laboratories for real-time analysis of seafloor sediments. DeMenocal took full advantage. He pored over the cores as they were raised from the ocean bottom. Each had alternating layers—creamy-coffee-colored sediments filled with the tiny shells of single-celled organisms called foraminifera, followed by darker, greenish-brown ones rich in continental dust. The pattern repeated over and over again, at fixed intervals each representing millennia of accumulation. This was the rhythm of the dance between Earth and Sun that Milanković had identified six decades earlier. But what struck deMenocal was something else—the dust-rich layers became thinner and lighter in color as the team drilled deeper into the seabed. He was not just seeing the effects of African monsoons on the accumulation of ocean-floor sediments, but he was also witnessing profound changes in the pattern as they pushed further into the distant past. By the time the *Resolution* pulled into Port Louis, Mauritius, after 56 days and 4400 nautical miles at sea, he had teased from those cores the basic relationship between orbital precession and climate in eastern Africa during the course of human evolution.

The details, though, would await deMenocal's return to Columbia University. Over the next year, he matched the cores with those recovered from an earlier leg of the ocean drilling project in the Atlantic off the coast of West Africa. A couple of things stood out. First, there was a connection between ice age events in the northern hemisphere and climate in the tropics and subtropics of Africa. At 2.8 mya, the pattern of alternating light and dark bands switched from 19,000–23,000-year cycles to one 41,000 years long—the same as the glacial-interglacial interval. After 1 mya, it shifted to a 100,000-year cycle, just as the northern hemisphere glaciers intensified.

OUR RESTLESS PLANET

Ocean-bottom sediments give us a very detailed picture of climate change on a broad geographic scale. If we're interested in the driving forces behind human evolution, though, we've got to understand local variation in habitats and resources where our hominin ancestors lived. Few places in Africa have changed as much as the Sahara has over the past 5000 years, and climate variation in the past must have also affected different places in different ways. So, many factors contribute to how a place responds to climate change: distance from the sea and equator, direction and intensity of prevailing ocean and wind currents, the shape of the land. And over the time scale of human evolution, these factors themselves change. This makes it even more important to think locally when considering relationships between climate change and evolution.

Let's consider tectonics. The Earth is a restless planet. Its outer shell is a patchwork of plates, massive slabs of rock that each drift over the underlying mantle. Movements of these plates completely change landscapes and seascapes at the boundaries between them. Some slide past along one another, forming a crack in the Earth's crust where they meet. The San Andreas Fault near the coast of California is an example. Some converge, with one pushed under the other to form a trench, like the Marianas in the Pacific. Colliding plates can also both push upward where they meet, forming mountain ranges, like the Himalayas that separate the Indian subcontinent and Tibetan Plateau. You can see the plate boundary clearly on images taken from space. And some plates diverge and separate, with new crust formed between them as magma pushes upward. The Mid-Atlantic Ridge, for example, is spreading at about an inch a year. It has a deep rift valley between its flanks, about the width and depth of the Grand Canyon.

How do these tectonic movements affect climate? First, shifting continents can open and close ocean corridors and alter patterns of circulation of water, salt, and heat. These are the engines of climate change. Remember from chapter 2 how even a slight change in temperature and air surface pressure in the tropical eastern Pacific withers the forest at Kibale in El Niño years? It should come as no surprise, then, that the opening of the Drake Passageway between South America and Antarctica and the closure of the Panama Seaway between North and South America during the Cenozoic had tremendous and far-reaching effects on climate. They set the stage for massive ice sheets to form over the south pole, and for the Gulf Stream to flow and moderate temperatures across the Atlantic, from Florida to Norway.[9]

Changes in landscape shape caused by tectonics are also important. When a range of tall mountains blocks the prevailing path of air currents, it causes a rain shadow on the leeward side. This is because atmospheric pressure declines with increasing altitude, so air expands and drops its moisture on the windward side. Death Valley in the western United States is in the rain shadow of the Sierra Nevada range. The Tibetan Plateau is in the rain shadow of the Himalayas. And the savannas and deserts of eastern Africa are in the shadow of the mountainous shoulders of the East African Rift System. Let's have a look.

The East African Rift System

We were introduced to the East African Rift System in chapter 2. Recall the gorillas in the cloud forests shrouding the Virunga volcanoes flanking its western arm, and the monkeys and chimpanzees at Kibale, on the edge of the Rwenzori Range between Lake George and Lake Albert. Those volcanoes and lakes are important parts of the system. The Great Rift Valley itself is actually a series of rifts running north–south between Syria and Mozambique, 2800 miles from end to end. These are the stunning, varied landscapes that Richard Burton and John Speke explored as they set out to find the source of the Nile—peaks and valleys, plateaus and basins. There are snow-capped mountains rising up to more than 19,000 feet above sea level and searing salt flats 500 feet below it. There are lush forests and parched deserts.

But it wasn't always that way. Geologists have only recently worked out the details. The region was much flatter during the middle part of the Cenozoic, and it was covered in tropical mixed forest. But a huge plume of magma was brewing far beneath the continent, and a massive upwelling of molten rock would soon rise and split the African tectonic plate in two. Volcanoes formed at the surface, and the crust bulged upward into a dome rising three-quarters of a mile in elevation. At the same time, a giant crack spread north to south, dividing in two around today's Lake Victoria, which overlies a patch of especially hard, ancient rock. As the new plate edges began to separate, tension caused the land above to subside and buckle, leaving behind the rift valleys and their raised shoulders.

Raymond Dart and the other early paleoanthropologists thought very little about the Great Rift Valley. They believed that our apish ancestors had travelled from central African rainforest south through the arid Kalahari to the South African veld. Recall from chapter 3 that Dart felt the trip itself had transformed our ancestors, as no mere ape could have conquered the challenges posed by such a treacherous, forbidding landscape. So human evolution was at first inextricably linked with southern Africa, or, at least, the journey to it. For Raymond Dart and Robert Broom, climate change played no role in human evolution because, as far as Dart knew, Africa had remained as it was since dinosaurs roamed the continent.

But it wasn't long before their successors realized that southern Africa *had* changed over time. We couldn't just assume that hominins were cast from their central African Eden to wander the great desert before entering the Promised Land. This, unbeknownst to researchers at the time, untethered human evolution from southern Africa and set the stage for the shift in attention to the East African Rift System. As we learned in chapter 3, Mary Leakey discovered *"Zinjanthropus"* in 1959, 1500 miles to the northeast at Olduvai Gorge in Tanzania. The eastern African fossil "gold rush" that followed produced hundreds of hominin specimens in a few short years, each recovered from deposits along Africa's Great Rift Valley in Tanzania, Ethiopia, and Kenya.

Not only did the Great Rift Valley yield evidence for human evolution, it also provided both motive and opportunity. A Dutch ecologist named Adriaan Kortlandt developed the basic storyline in the early 1970s. Today, moist air from the Atlantic blows eastward, but the peaks that line the western rim of the Great Rift Valley form a barrier of sorts to limit rains as they move across the continent. This explains the stark contrast between the lush Congo Basin and arid eastern Africa. Kortlandt proposed that those ridges rose during the course of human evolution, casting their rain shadow, withering forests to the east, and trapping our ancestors in the savanna that remained. Yves Coppens, one of the leaders of the original Omo Expeditions, laid out the argument in his 1994 *Scientific American* piece "East Side Story."[10] The idea was that the ancestors of humans and chimpanzees became separated. Those to the west were adapted to wet forests and evolved into the chimpanzees, whereas those to the east faced an increasingly dry, open landscape. The take-home message was clear. Our early ancestors were not cast out from the Garden of Eden. They were not transformed by leaving their ancestral home, as Dart had reasoned. Instead, their world changed around them, and the protohominins were forced to change along with it.

Hominin Habitats

The East Side Story is very compelling. There's evidence, motive, and opportunity. In truth, though, we have no idea whether the spreading Great Rift Valley made us human. Yes, eastern African hominins from younger deposits look progressively more humanlike. And yes, there are plenty of environmental changes that could have triggered their evolution. But other places changed too, and early hominins have been found at sites from the Djurab Desert in Chad to the Bloubank River Valley in South Africa. Our early ancestors may well have ranged across much of the continent during the Plio-Pleistocene, facing all sorts of environmental challenges and opportunities. But we've got to start somewhere, and eastern Africa makes a lot of sense. Eastern Africa is a principal focus of attention for paleoecologists today, and it gives us the detail we need to begin matching paleoenvironment with the hominin fossil record.

The Great Rift Valley has drawn many gifted geologists and paleontologists over the past half century. It's provided an exciting, stimulating atmosphere for them to develop and experiment with new tools to
tease out information about the past. Thure Cerling was among the first
in, with an innovative and groundbreaking approach to reconstructing
past habitats—the chemical composition of the fossil soils in which the
early hominins were buried.

Cerling was an undergraduate majoring in geology and chemistry at
Iowa State University in the late 1960s and early 1970s. The geology
department there at the time was on the rise, with a fresh infusion of
creative and energetic young scholars like Carl Vondra, who studied
the accumulation of sediments at fossil hominin sites. And the chemistry department was keen on students double majoring. Harry Svec,
an alumnus of the Manhattan Project, was at the time pioneering new
approaches to combining chemistry and geology. Cerling took full advantage of the opportunities that Iowa State had to offer and managed
to score an invitation to help map sediments on the eastern shore of
Kenya's Lake Rudolf (now Lake Turkana). The hominin fossil rush in
eastern Africa was starting to heat up, and there was much to do to understand the geology of the new sites. What he learned in those early
years about geochemistry and tracing deposits across the landscape
would serve him well into the future.

Cerling continued to work in Kenya through the 1970s during his
graduate school years at Berkeley. His passion was the chemistry of
sediments, and he began to collect the popcorn-like calcium carbonate
nodules that accumulated over time at the sites. He spent months each
year walking the deposits, gathering samples from each of the layers
that had produced fossil hominins. It was blistering work, tracing layers
of rock and sediment mile after mile, day after day, across the hot and
dusty badlands on the eastern edge of the lake. The search took him
up, down, and around erosional gullies and barren hills broken only
by the occasional bush or acacia tree. I've always been jealous of geologists who can look out over a desolate landscape and see the past.
Rocks and dirt become ancient streambeds, river deltas, and lake shore-
lines in their mind's eye. Cerling knew exactly where to sample.

His tools were pickaxe and shovel, and he had to dig deep enough to
reach the precious pristine nodules, unexposed to air and the elements.

4.2. Thure Cerling collects calcium carbonate nodules for isotope analyses. Image courtesy of Thure Cerling.

Cerling collected hundreds of samples, not knowing for sure whether they were contaminated until he could get them back to the lab. But it was well worth the effort. Geochemists at the time were starting to realize that there was much to be learned from the chemistry of fossil soils. Let's consider variants, or isotopes, of oxygen and carbon atoms. Cerling's calcium carbonate nodules ($CaCO_3$) had both. More than 99% of all oxygen atoms in our atmosphere have eight protons and eight neutrons (^{16}O), and most of the rest have two extra neutrons (^{18}O). Also, about 99% of carbon atoms on Earth have six protons and six neutrons (^{12}C). Most of the rest of those have a seventh neutron (^{13}C). The ratios of carbon and oxygen of each type are reasonably constant in the atmosphere, but they weren't in Cerling's calcium carbonate samples. In fact, individual geological layers seemed to have nodules with distinctive isotopic signatures. Some had more heavy oxygen (^{18}O) or carbon (^{13}C) than others. This would help him match up and trace deposits across a landscape.

Cerling also realized that the differences in isotope ratios must have been caused by differences in soil conditions when the deposits were formed. Maybe, then, he thought, chemistry could offer clues about past environments to help us better understand the challenges our ancestors faced during their evolution. For example, there is more heavy ^{18}O in rain at high temperatures than at low ones, because cooler clouds hold less ^{18}O. So water found in warmer soils should have a higher proportion of $H_2^{18}O$, all else being equal.[11] In other words, oxygen isotope ratios in calcium carbonate should be able to teach us whether conditions were warm or cool when the sediments were laid down.

Carbon isotopes are also helpful. Carbon deep in the soil comes from carbon dioxide produced by plants, animals, and bacteria. Different kinds of plants (along with those that eat them) have different proportions of carbon atoms with six and seven neutrons, depending on how they "do" photosynthesis. Trees, shrubs, and cool-season grasses, for example, have lower proportions of heavier ^{13}C atoms than do low-altitude tropical grasses and sedges.[12] So fossil soils formed on what were tropical savannas should have higher ratios of ^{13}C to ^{12}C atoms than those that supported woodlands or forests. It should be possible, then, to use this to infer something about the habitats in which our ancestors lived and evolved.

But soil isotopes aren't easy to interpret. Temperature, sunlight, humidity, vegetation cover, depth below the surface, and other things affect the ratios. Cerling would have to look to modern soils to work out the details. So he and his student, Jay Quade, spent much of the 1980s and 1990s collecting and analyzing soils from the plains, prairies, and deserts of North America; the equatorial forests and grasslands of South America and Africa; the woodlands and shrublands of Mediterranean Europe and Australia; and the grassy floodplains of Pakistan. It was a monumental undertaking that took years of study but, in the end, the two managed to free the environmental clues trapped in calcium carbonate, to wring temperature, precipitation, and vegetation type from the oxygen and carbon in ancient sediments.

Now Cerling could turn back to the fossil carbonates from the hominin sites. He expected the carbon isotopes to show increasing proportions of heavy carbon in younger deposits, indicating the spread of tropical grasses from the late Miocene on, and, at the same time, less heavy oxygen, suggesting increasingly arid conditions. And indeed, there were trends. It looked as though tropical grasses and sedges had begun to develop and expand into the floodplains of the Great Rift Valley by the middle to late Pliocene. There was an increase in heavy carbon in deposits dated to about 1.8 mya, consistent with widespread wooded grasslands. At the same time, heavy oxygen atoms dropped off, suggesting drier conditions. Cerling was witnessing the expansion of savannas across the east side of the continent. It seemed to be happening at about the same time hominins were stepping down from the trees, and grass-grazing antelopes were replacing forest and woodland browsers, just as earlier researchers had predicted (see chapter 3).

But it wasn't that simple. Yes, there were general trends through the Plio-Pleistocene, but the timing was different in different places. It had become drier earlier along the Awash River in central Ethiopia than farther south where the Omo River feeds Lake Turkana near the border with Kenya. There were differences within areas too. Woodlands persisted longer along the Omo than they did on the shoreline of the lake. A quick check on other parts of the world confirmed differences across the globe—heavy carbon grasses have dominated the Siwalik Hills of northern Pakistan, for example, for the past six million years.

But what about those fine-scale fluctuations over time that Peter deMenocal had identified? How did Milankovitch cycles and the spreading Rift Valley work in sync on the local level to produce changes in conditions that individual hominins had to face? That is, after all, the level at which natural selection operates. How did the landscape vary from one generation to the next, and how did those who lived there cope with these changes? Let's head over to Olorgesailie in Kenya and visit with Rick Potts of the Smithsonian Institution's Museum of Natural History to look for answers.

FOCUSING IN ON THE FLUCTUATION

Olorgesailie is about an hour's drive southwest of Nairobi along the narrow, pothole-laden road to Lake Magadi. You descend 2000 feet from the Ngong Hills to the Rift Valley floor. It's a hot and arid place, an open plain dominated by grasses, thorny bushes, and the occasional acacia tree. But it's easy to envision a lake filling the basin in your mind's eye, its shoreline teeming with life.

The geology tells the story—layer upon layer of sediment, altogether more than 250 feet thick. The layers stretch out across an area about 60 square miles—white, green, brown, red, gray. These are the telltale colors of lake beds, age-old soils, river channels, wildfires, and volcanic ash falls. Geologists use them to piece together the changing landscape between about 1.2 mya and 500 kya. The white layers, for example, are filled with the shells of diatoms, algae that settled on the lake's floor back when Olorgesailie was a wetter, more hospitable place. We can learn a lot about the lake—how deep, salty, and acidic it was—by the types of diatoms found in those layers. The changing deposits over time and space are like a window into the basin's past.

Researchers have been working at Olorgesailie for a long time. Stone hand axes were first reported there by a British geologist, John Walter Gregory, in 1919. He discovered them during a walking safari from Nairobi to Lake Magadi. Louis and Mary Leakey went there in 1942 and found hundreds more eroding from the sediments. You can still see many of them today, from a raised wooden walkway at the site, jutting out of the ground. Mary Leakey and her team found plenty of fossils during the war years too, though no hominins.

The first hominin wasn't found at Olorgesailie until more than half a century later. It was a *Homo erectus* skull discovered by Rick Potts in 2003. Potts had been working at Olorgesailie since 1985. Back then archaeologists were beginning to look beyond concentrations of stone tools and bones, and to think about how hominins might have used whole landscapes. Olorgesailie was a huge site, with deposits stretching as far as the eye could see, and it was a great place to look for what archaeologist Glynn Isaac,[13] who had excavated there in the 1960s, called "the scatter between the patches." So Potts and his team got to work digging trenches, 100 of them, across a span of three miles of outcrop.

Potts at first figured that he'd work the site for about three field seasons. But as he spent more time there, he became drawn more to what was happening over time than across space. Olorgesailie has more than 700,000 years of stacked sediments, and the animals found in earlier deposits were different from those in the later ones. This in itself wasn't surprising. Vrba's idea that pulses of extinction and evolution followed the savannas as they spread across the continent (see chapter 3) was at the peak of its popularity. But she had focused on earlier times, the boundaries between the Miocene and Pliocene around 5.3 mya, and between the Pliocene and Pleistocene at 2.6 mya. Vrba thought there might also have been a turnover around 1.0 mya, but the particulars were sketchy. Olorgesailie was in the right place at the right time to fill in the details.

But the fossils coming out of the ground at Olorgesailie didn't make much sense in light of Vrba's turnover-pulse hypothesis. Rather than open-country species replacing woodland or forest-dwelling ones, Potts found grass specialists, like geladas and hippos, in the lower deposits, and mixed feeders, like baboons and elephants, above them. The latter species could earn a living in either savanna or forest. It was as if dietary specialists at Olorgesailie were replaced by flexible generalists. Could this teach us something about the hominins who made and dropped their hand axes there?

Potts began to wonder whether those things that make us human, like bipedalism, big brains, complex social organization, stone tools, and meat eating, might have evolved to give our ancestors the versatility they needed to thrive in a broad range of habitats. Maybe the key to our evolution wasn't spreading grasslands, but the uncertainty that came with ever-more fluctuating environmental conditions and the challenges they pose. He began combing the literature on paleoclimate to better understand those fluctuations. By the mid-1990s, Potts had pieced together the variability selection hypothesis.

DeMenocal, Cerling, and others had documented fluctuating temperature and precipitation superimposed on a trend toward cooler, drier conditions. But Potts needed to know how the oscillations themselves had changed over time. Was there evidence for greater swings during key periods in human evolution? It wasn't clear from oceanic dust or soil carbonates at the hominin sites. But there were other climate proxies—the oxygen isotope record for fossil shells of foraminifera, for example. These accumulate on the seafloor over millions of years, and can be used as a sort of deep-time thermometer to measure changes in temperature. A relatively higher ^{18}O level in foraminifera shell indicates that more ^{16}O was bound up in glacial ice at the time, and average global temperature was cooler.[14]

Nicholas Shackleton at Cambridge and his colleagues had just assembled a dataset of foraminifera oxygen isotope values from cores

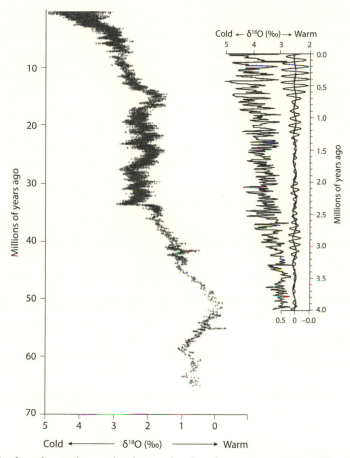

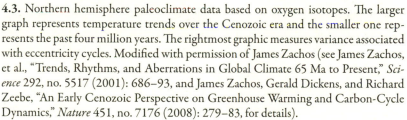

4.3. Northern hemisphere paleoclimate data based on oxygen isotopes. The larger graph represents temperature trends over the Cenozoic era and the smaller one represents the past four million years. The rightmost graphic measures variance associated with eccentricity cycles. Modified with permission of James Zachos (see James Zachos, et al., "Trends, Rhythms, and Aberrations in Global Climate 65 Ma to Present," *Science* 292, no. 5517 (2001): 686–93, and James Zachos, Gerald Dickens, and Richard Zeebe, "An Early Cenozoic Perspective on Greenhouse Warming and Carbon-Cycle Dynamics," *Nature* 451, no. 7176 (2008): 279–83, for details).

of deep-sea sediments accumulated over the past six million years. They lined them up, plotted them out, and found a striking increase in amplitude of oscillations over time—a two- to threefold rise during the course of hominin evolution. Here, perhaps, was Potts's evidence. There were other hints too. Pollen from a peat bog deposit in northern

Greece, for example, showed an increase in the intensity of fluctuating conditions there, back and forth between temperate forest and cold, open steppe. Climate change, it seemed, wasn't just about long-term trends or regular, short-term oscillations. Intensity of those fluctuations also had to be considered. Maybe it wasn't long-term pulses of aridity that drove our evolution at all but, rather, short periods of high climate variability. Perhaps our ancestors became increasing adaptable, and that allowed them to keep coming back to Olorgesailie, when so many others did not.

What about the Lakes?

There are many parts to the story of our changing world. Regular variation in the Earth's orbit and axial tilt set the pace for climate fluctuation on a global scale. Tectonic activity can help us understand environmental changes on the regional and local levels. But what's the connection between the parts? Mark Maslin at University College London and Martin Trauth at the University of Potsdam in Germany might just have the answer. They say it's the lakes.

Maslin was a student of Shackleton's back in the late 1980s and early 1990s, and had worked with him on oxygen isotopes of foraminifera in deep-sea cores from the ocean drilling program. As Peter deMenocal did before him, Maslin cut his teeth on the *JOIDES Resolution*. He was a postdoc at the Institute of Geology and Paleontology at the University of Kiel in Germany at the time. That leg (number 155 in 1994) explored the huge fan of mud dumped by the Amazon River into the Atlantic Ocean off the coast of South America. While Maslin was at Kiel, he met Trauth, who was developing mathematical models for climate change. Years later, Trauth invited Maslin to join him for some geological work in Kenya. The two began working together and before long had an ambitious project to document changes in East African lakes over time.

The parting of Africa has left basins all up and down the Great Rift Valley. Many have filled with water during humid periods and emptied during arid ones. These expand and contract following Milankovitch cycles. Others had already found bits and pieces of evidence for this. Fossil diatoms from deposits between 2.7 and 2.5 mya in Kenya's

Baringo Basin showed a series of lakes, one after the other, coming and going at about 23,000-year intervals, following the precession of the equinoxes (summer at our closest approach to the Sun, then winter, then summer again). Maslin and Trauth looked across 10 major lake basins to build on this work. They found a consistent pattern again and again. Major lakes came and went at 400,000-year intervals, at least until the Pleistocene,[15] following the cycle of eccentricity of the Earth's orbit from near circular to oval and back. And the lakes were deepest during important periods of global climate change.

Rift Valley lakes are especially sensitive to small changes in rainfall. Humid periods mean not only more precipitation, but also less evaporation. Arid times mean the opposite. This amplifies the effects of Milankovitch cycles. Climate swings come fast and furious, and rainfall levels quickly reach thresholds for expansion and contraction of the lakes. In other words, basins fill and empty in pulses.

This sets up Maslin and Trauth's pulsed climate variability hypothesis. It builds on Potts's idea that environmental fluctuations are key to human origins, uses lakes as a sort of bridge between the effects of orbital dynamics and tectonics, and helps explain why some evolutionary events seem to have occurred abruptly rather than gradually over time. The idea is that as lakes fill and empty, they disrupt life in the rift basins. When conditions get too wet, the basins fill and hominin populations split and move up and along the rift shoulders. The basins also become uninhabitable when they get too arid, so hominins again split and move. As we learned in chapter 3 from Elisabeth Vrba, this kind of fragmentation and dispersal of populations is the stuff of evolution. Perhaps it explains how changes in the Earth's orbit and tilt, combined with the splitting of Africa, helped make us human.

Down into the Future

So where's the evidence? We don't yet have the archive of long-term, fine-scale climate changes in the right places and at the right times to know just how they triggered specific events in our evolution. Ocean cores give us great continuous records over deep time, but not the detail on local environments that we need. Soil carbonates at hominin sites give us that detail, but not the continuous record. Deposition, exposure, and erosion of fossil soils limit what we can say about hominin environments to broken and disconnected bits and pieces of time.

But that's all starting to change. Researchers are just now beginning to probe into the ancient beds of the Rift Valley lakes themselves. Lake beds accumulate continuously over time, much like ocean-floor sediments. So they hold details about regional and local environments in the Great Rift Valley. Fine layers of sediment, diatoms, charcoal, plant-wax residues, and pollen build up, one atop the other, to give us a record that can be read just like the ocean cores. They chronicle changes in temperature, precipitation, and local vegetation across years, decades, centuries, and millennia. Bits of volcanic ash in a core can be dated, and individual layers matched with fossil hominins and other animals nearby. Lake cores can transform the way we do our science and help us answer questions that early paleoanthropologists could not have even dreamed of asking.

It's required years to plan and dogged determination to raise the funds. Drilling in eastern Africa is very expensive and a logistical nightmare. Imagine the effort it takes to move, set up, and run a 100-foot-high drill rig in the remote reaches of the Great Rift Valley. Think of the challenge of getting the water needed to cool the rig on such a hot and arid landscape. Lifting samples from boreholes requires very specialized equipment, and when something breaks, and it often does, you can't just run down to the local hardware store for a replacement part. It takes all the ingenuity, patience, and sense of humor a scientist can muster to cope with the challenges.[16] But it's happening, and it's thrilling.

Rick Potts and his team drilled two boreholes at Olorgesailie in 2012. The cores they lifted total more than 700 feet and amount to half a million years of accumulation. Another team directed by Andrew Cohen at the University of Arizona has just completed a massive effort to drill boreholes in six lake beds scattered between northern Ethiopia and southern Kenya. The combined cores total more than a mile and sample key periods over the past 3.5 million years.[17] It's a multimillion-dollar project, with a team of 100 scientists from 11 countries. Those samples will keep the team busy for years to come, but their results

4.4. Olorgesailie drilling operation (*left*) and Rick Potts examining core samples (*right*). Images courtesy of Jennifer Clark, Human Origins Program, Smithsonian Institution.

promise to give us unprecedented details of climate change over deep time and how it shaped our ancestors' worlds.

CLOSING THOUGHTS

We don't yet have all the details, but some things are already clear. The environments that early hominins lived in varied a lot over the course of our evolution, and changes to the Great Rift Valley and elsewhere in Africa must have had a dramatic effect on those who lived there. Not only are many parts of the continent more arid than they were 5 or even 2.5 mya, but hominin habitats must have flipped back and forth between warm and wet and cool and dry with global-level climate fluctuations. Sometimes the change was abrupt; other times, it was gradual. Each time a basin filled and emptied, nature must have cleared the biospheric buffet with a single swipe and restocked it with a new set of food choices. What effects did these changes have on our ancestors? Perhaps the teeth and bones of the fossils themselves hold some clues. It's time to head back to the lab.

CHAPTER 5

Foodprints

Louis and Mary Leakey had brought their newfound fossil skull to Léopoldville, now Kinshasa, for the Pan-African Congress on Prehistory in August of 1959. It was the first early hominin skull found outside of South Africa and they had discovered it just the previous month at Olduvai Gorge. Everyone at the meeting was talking about it. The cranium had a thick, sturdy upper jaw and face; a large bony crest that must have anchored massive chewing muscles; and enormous cheek teeth. Phillip Tobias, who would later write the definitive book on the specimen, said, now famously, "I have never seen a more remarkable set of nut crackers."[1]

"Nutcracker man" *was* remarkable. The species had the most powerful jaw and biggest, flattest teeth of any hominin discovered before or since. The skull looked so much like that treasured wooden toy in its bejeweled soldier uniform, walnut in mouth. What else but a steadfast hard-object feeder could it have been? It made perfect sense, too. Grasslands spread across eastern Africa in the early Pleistocene as the Great Rift Valley expanded and its shoulders rose, in sync with global climate cooling and drying. The fleshy fruits of lush forests became scarce, and hard, dry foods, like acacia seeds and tubers, more plentiful. It seemed almost inevitable that a specialized hominin with nutcracker-like chewing anatomy would evolve under those conditions. And this remained the prevailing view for nearly half a century.

That's why I was so surprised by what I saw when I first examined replicas of the skull's teeth with my new microscope. I saw nothing but wispy scratches. Hard-object feeding should cause pitting as foods are crushed between opposing teeth and tiny abrasive particles are pressed into their surfaces. My postdoc, Rob Scott, had already scanned repli-

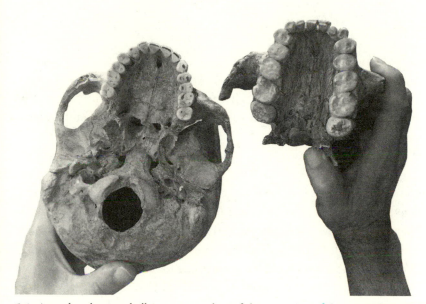

5.1. A modern human skull next to a replica of the upper jaw of the nutcracker man (*Paranthropus boisei*).

cas of the teeth of other early hominin species, but I had saved this one for myself. *Paranthropus boisei* was my favorite, and I wanted to be the first to see its battered biting surfaces in 3D, riddled with deep craters formed in life when the hominins crushed hard nuts, seeds, and roots. But the surface wasn't pitted. I looked at another tooth; again, there were wispy scratches. Specimen after specimen, it was the same. This was no nutcracker. Had we all gotten it wrong?

This chapter takes us into the laboratory for a closer look at the hominins themselves. When we consider their teeth in light of what we have learned about primate feeding ecology and environmental change over deep time, our reconstructions of diet based on tooth form seem naïve. We've worked out how different foods fracture and considered the best tools to break them, but that's not enough to know how our distant ancestors made use of the spreading savanna to earn a living. This is no simple engineering problem. We need to think like paleo-ecologists, to consider fossils, as George Gaylord Simpson sought to, "not as bits of broken bone, but as flesh and blood beings."[2] This requires a very different approach.

SOLVING FOR *X*

Astronomer Carl Sagan used to say, "You have to know the past to understand the present." Those of us who study the past think about things the other way around. There's a simple formula I write on the blackboard whenever I'm lecturing about the evolution of human diet:

$$\text{Living primate}\left(\frac{\text{teeth}}{\text{diet}}\right) = \text{fossil hominin}\left(\frac{\text{teeth}}{x}\right).$$

We've got to learn what living primates choose to sustain themselves in the forest, and why, to work out relationships between teeth and diet. We need to know not just how teeth work, but how animals use them today to solve for *x*. Remember lesson one from Ketambe in chapter 2. We can no more appreciate how teeth are used outside the context of the forest than we can understand how a heart or brain functions after cutting it out of the body.

Recall that Bob Sussman's brown lemurs and ringtails are able to live together in a single patch of forest because of the way they divvy up potential foods. Brown lemurs eat more leaves and ringtails eat more fruits. You'd never guess it from their teeth, though, because their shearing crests are about the same length. In that case, the differences in diet are not about the *types* of food eaten but, rather, their proportions. Both species eat fruits, leaves, flowers, and bark, and that's what seems to be important when it comes to selection for tooth form. That doesn't mean, of course, that the amount of each food they eat isn't just as critical to the roles the two species play in their world. It's just that you might not see it in the sizes or shapes of their teeth.

The same is true for gorillas of the Virungas, Bwindi, and Bai Hokou. Recall from chapter 2 that their diets vary by altitude. Higher elevation means more wild celery because favored fleshy fruits are less plentiful. Diet tracks seasonal availability of fruits at a given place too. But they all have the same sharp, crested teeth for shredding tough, fibrous plant parts. We can't tell from the shapes of their teeth that gorillas differ in their diets at different times and in different places, let alone that they prefer soft, sugary fruits when they can get them. The mangabeys at Kibale give us another example. They too prefer fruits, despite powerful jaws and thick, flat teeth suited to crushing the hard bark and seeds they have to eat to survive the lean times when preferred foods are scarce.

The relationship between tooth shape and diet at first seems reasonably straightforward. Remember the lion and giraffe teeth in Little Rock's Museum of Discovery back in chapter 1? It's pretty obvious, even to small children, that sharp teeth are for meat and blunt, ridged ones are for plants. But is it really that simple? Can we really use tooth shape to reconstruct the food preferences of fossil species? Not if those species acted like the gorillas at Bai Hokou or the mangabeys at Kibale, which actually avoid the very foods to which their teeth are adapted when given a choice. This in fact happens a lot in nature. Consider Minckley's cichlid, a freshwater fish from Mexico and southern Texas. One form of this fish has flat, pebble-like teeth perfectly adapted to crack hard snail shells. But, as the late ichthyologist Karel Liem noted, they pass right by snails whenever there are softer foods available.

It may seem counterintuitive to evolve teeth and jaws for less-favored, rarely eaten foods, and paradoxical that more-specialized anatomy can lead to a more-generalized diet, but there are many, many examples of this in nature. We call the phenomenon *Liem's paradox* in honor of Karel Liem. So long as hard-object adaptations, for example, do not preclude consumption of softer items, they can actually lead to a longer menu of potential foods. Think of it this way. It doesn't matter much what your teeth look like if you eat mush 360 days a year, but if you have to eat rocks to survive the other 5, you'd better have teeth that can crush rocks.

In some cases, though, specialized anatomy does reflect a specialized diet. Recall the Taï Forest mangabeys. They use their powerful jaws and thick, flat teeth to eat hard nuts from the forest floor nearly every day. This is part of a strategy for divvying up the forest's bounty with other primates. The sooty mangabeys take the ground, and other monkeys at Taï take the trees. So, again, morphology is about kind, not proportion. Nature responds to the most challenging foods a species has to contend with to survive, whether eaten daily or just often enough to endure the hardest or toughest of times.

The take-home message is that tooth size and shape aren't enough for us to understand how our ancestors responded to changing food availabilities over the course of our evolution. Yes, these things can teach us something about what a species was capable of eating, and likely the selective pressures that led to the shapes of their teeth today. But what about the foods a specific animal actually ate during its lifetime? What about its preferences and its daily choices? These questions are just as important, but we need a different approach to address them. We need what I call *foodprints*, traces of diet that leave telltale clues on or in the teeth when an animal eats. These traces can be the distinctive patterns of microwear—microscopic scratches and pits left by foods that are hard or soft, brittle or tough. Or they can be chemical signatures in teeth that came from the foods that supplied the raw materials used to build them. Like footprints in the sand, foodprints give us evidence of actual activities of real animals at a moment in time in the past.

This is a profoundly different way of looking at fossils than the traditional "sharp teeth are for meat and blunt ones are for plants" approach. The focus shifts from theory and models for what might have caused traits to evolve in a species to real foods eaten by the very animal whose fossilized teeth we hold in our hands. It becomes more than an issue of what a species could have eaten and what was available to it on the biospheric buffet. It becomes one of what an individual in the past actually selected to put on its plate.

DENTAL MICROWEAR

It's easy to see how chewing can cause microscopic scratching and pitting of teeth, and it makes sense that different patterns should come with different diets. But working out the particulars so we can translate microwear on fossil teeth into diets of long-dead animals has been a real challenge. There are so many variables to account for. Animals vary in the shapes of their teeth and they chew in different ways. They live in different habitats and are exposed to different amounts of abrasive dust and grit. How can we even know that microwear reflects diet at all?

From Hyraxes

This question was very much on Alan Walker's mind back in the early 1970s. Walker was my first postdoctoral adviser, himself a product of John Napier's primate research unit at the Royal Free Hospital School of Medicine in London, as were Colin Groves and Cliff Jolly (see chapter 3). He had gone on to teach anatomy in Africa, first in Uganda, then Kenya. It was in Kenya that he first saw images of cells produced using a scanning electron microscope, or SEM. A pioneer of electron microscopy was visiting from Harvard and gave a talk at the University of Nairobi. The resolution, clarity, and depth of field were amazing, like nothing Walker had ever seen before. Think of those vivid black-and-white pictures of sperm meeting egg in your high-school biology textbook. If the SEM could produce such remarkable imagery of living cells, surely it could open whole new avenues for studies of fossils.

Walker thought about tooth enamel. Work on lemurs and lorises half a century earlier had suggested that more closely related species had more similar enamel. Could an SEM help him work out relatedness of fossil species from subtle variations in the microscopic structure of their teeth? He arranged for some time on the new instrument at Cambridge during his next leave from Nairobi and, armed with some lemur and loris teeth, set out to see. He didn't find the differences he was hoping for.

But he did notice something else—crisp scratches and pits carved into the wear surfaces of the teeth. Surely those were related to chewing. It seemed that the SEM couldn't help him work out relatedness, but maybe it could provide insights into diet. Walker began to develop a plan once he got back to Kenya. He would need closely related living species with teeth of similar size and shape to be sure that differences in dental form wouldn't affect the pattern. They would have to inhabit the same environment so he could be sure differences in abrasive soil or dust level wouldn't affect the pattern. And they would have to eat different things. But where could he find the right animals for the study? He described the idea to Henrick Hoeck over a beer in Nairobi. Hoeck was visiting from the Serengeti Research Institute, and he had just the right sample.

Hoeck had compared the diets of two species of hyrax. These small, furry animals are rather odd. They resemble marmots or woodchucks, but are more closely related to elephants and manatees. The bush hyrax and rock hyrax live together on rocky outcrops in the Serengeti and share the same crevices and cavities. They are difficult to tell apart to the untrained eye, though the rock hyrax is slightly larger and has a longer snout. More to the point, they have very different diets, especially in the rainy season. The bush hyrax is a browser. It mostly eats bits of bushes and trees. The rock hyrax is a grazer. It eats more grass, though in the dry season grasses become parched and the rock hyrax shifts to more bush and tree parts. Hoeck had collected some specimens in the Serengeti and invited Walker to study their teeth.

Walker soon after moved to begin work at Harvard and put high-resolution replicas of Hoeck's hyrax teeth under an SEM there. Differences between the species were obvious. The bush hyrax had smooth, polished teeth, but the rock hyrax, at least individuals collected during the rainy season, had enamel covered in fine, parallel scratches. And those differences made sense. Grasses often contain bits of silica called phytoliths. The silica is leached from the soil, absorbed through the roots with water, and accumulated in and around a plant's cells. Walker reckoned that the scratches must have come from phytoliths in the grasses that the animals were chewing. He found phytoliths, many fractured, in fecal pellet samples from the rock species but not the bush species. He also found that the teeth of rock hyraxes collected in the dry season were polished like those of bush hyraxes, just as one might expect. The microwear tracked seasonal change in diet! Walker had discovered an important new tool for reconstructing feeding strategies of individual animals in life from their teeth. This was big! He published his study in the journal *Science* in 1978.[3]

To Hominins

But Walker was not alone. As scanning electron microscopes found their way into university laboratories across the globe in the 1970s, other researchers interested in fossil teeth also discovered them and realized their potential for microwear research. My PhD adviser, Fred Grine, used the new SEM at the University of the Witwatersrand to document wear on the teeth of mammal-like reptiles that lived 240 mya. He had gone to Wits to study human evolution with Phillip Tobias back in 1975, but soon found himself volunteering on fossil expeditions to the Karoo in search of much more ancient mammal-like reptiles. *Diademodon* in particular had sparked his interest. It was a bulky animal, about the size and shape of a razorback hog. It had long, piercing canine teeth and back teeth varying from simple cones to more ornate structures with rows of small cusps.

Diademodon was one of the first animals with mammal-like jaw muscles and precise occlusion between the upper and lower teeth. Did it chew like mammals, with horizontal movements of the lower jaw, or were there only vertical movements, like most living reptiles? This was an important issue for those interested in the origin of mammalian chewing and, at the time, a matter of some debate. Grine reckoned that he could figure it out with the help of the SEM at Wits. He reasoned that horizontal movements of the teeth should produce scratches as opposing surfaces slid past one another, but that pits should form if lowers merely pressed upward against the uppers. Grine found only pits. *Diademodon* had not, it seemed, developed mammal-like chewing after all.

But he had gone to Wits to work on early hominins, and Tobias was beginning to wonder whether he was losing interest. Grine needed a project to prove otherwise. This was just about the time Milford Wolpoff at the University of Michigan developed a provocative new theory suggesting that the hominins *Paranthropus* and *Australopithecus* were really male and female of a single species. How could two closely related, culture-bearing species coexist without one competing the other into extinction?[4] But the sizes and shapes of their teeth *were* different, and John Robinson, Cliff Jolly, and others had made a compelling case that those differences meant different diets (see chapter 3). If their diets were different, Grine reasoned, they must have been different species.

Fred Grine had his project. If *Australopithecus* and *Paranthropus* ate different foods, their teeth should evince different patterns of microscopic scratches and pits. He would use the SEM at Wits to see, just as he had done for *Diademodon*. While his first study involved fewer than a dozen milk teeth, he found what seemed to be differences. It wasn't at all clear to him what they meant at first, because there was nothing to compare them with. But Walker's work published on hyraxes the following year gave Grine a place to start. He looked at more teeth, and began documenting microwear patterns in earnest. *Australopithecus* molars were covered in microscopic scratches, whereas those of *Paranthropus* were more heavily pitted. These hominins ate different things, and so they were different species. It looked like *Paranthropus* crushed its foods and so ate harder or more fibrous ones than did *Australopithecus*.

Working Out the Details

It was clear heading into the 1980s that microwear had promise, but that the devil was in the details. Alan Walker had moved once again, this time to work at Johns Hopkins, and enlisted the help of a couple of young anatomy instructors, Kathleen Gordon and Mark Teaford, to help work them out. Gordon focused on methods. What were the best settings for the SEM? How could scratches and pits be measured? Which tooth in the row, and which wear facets on that tooth, should be studied? All of these things made a difference and needed to be taken into account. Teaford began looking for differences in microwear between primate species. He made many trips, armed with his trusty dental impression material dispenser gun, to the Smithsonian Institution's Museum of Natural History in nearby Washington, DC, to slather ever more specimens with tooth-molding goo.

Teaford was able to match the microwear pattern for individual species with diets documented by primatologists in the field. The Smithsonian had some gorilla skulls collected at Dian Fossey's Karisoke Research Center in Rwanda (see chapter 2). Those wild-celery-eating gorillas, along with leaf-eating howling monkeys in the collection, had microwear surfaces covered in fine, parallel scratches. Brown capuchin monkeys and mangabey species known to eat hard foods—like nuts, palm fronds, and bark—had fewer scratches, but more large, deep pits. Fruit-eating orangutans and chimpanzees had intermediate microwear patterns. Those patterns made sense. If tough leaves and stems require shearing, abrasives in or on them should be dragged along the surfaces of opposing teeth as they slide past one another. That should result in scratches aligned in the direction of tooth movement. And if hard objects are crushed, pits should form as they are pressed between the enamel surfaces of upper and lower teeth. The microwear patterns were just as Teaford and Walker expected, given differences in diet and their understanding of how teeth break down foods with different properties.

Fantastic Voyage

My own first experience with microwear came while Walker and his team were in the midst of their studies. I was a first-year graduate student at Stony Brook University, and Fred Grine took me to the Brookhaven National Laboratory, just a few miles east of campus. The SEM there had a seemingly unmanageable maze of knobs and lighted buttons, a couple of old TV-style screens, and a large vacuum chamber that housed the gun that fired the electrons and the detector that read the signal. We put our precious, gold-plated tooth replicas on the stage, closed the chamber, and pumped out the air. Grine worked the knobs and buttons like an expert pilot, and we watched the grainy images on the screen as he tilted, rotated, and zoomed in on the wear surfaces of each specimen. It was like magic.

I spent a lot of time on SEMs after that. I relished the feeling of solitude during hours spent in those small, darkened rooms, staring at screens glowing in iridescent shades of green, amber, or gray. I was carried away. The SEM became my ship, and its screens were my portholes to a microscopic world. I explored vast tooth surfaces, diving into narrow canyons and crossing enormous basins. But there was more to it than that. Those scratches and pits were also my windows on the past. They were formed in the mouth of a real flesh-and-blood being, maybe a great-times-100,000 grandparent, aunt, or uncle. With each new specimen, I became witness to an event that happened millions of years ago.

Deciphering the Past

Microwear scratches and pits are like ancient inscriptions engraved in teeth and left behind by our ancestors. My mentors—Alan Walker, Fred Grine, and Mark Teaford—had developed an elaborate method for deciphering those inscriptions. First, it required replicas. Original hominin fossils are too rare and precious to risk borrowing them from museums, and SEMs at the time were too big to bring to the fossils. It takes incredible resolution to duplicate tiny scratches less than 1000th the width of a human hair. Each tooth had to be cleaned carefully, because even the thinnest layer of dust or fingerprint grease blankets a

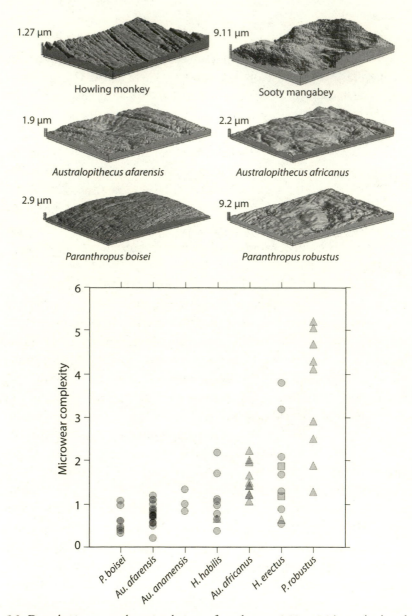

5.2. Dental microwear photosimulations of tooth areas 0.10 × 0.14 mm (*top*) and microwear texture complexity data for Plio-Pleistocene hominins (*bottom*). Eastern African hominin specimens are represented by gray circles, South African ones by triangles, and Asian ones by squares.

microwear surface and renders a replica useless. Then a special vinyl silicone was used for molding—the goo described earlier. The detail preserved in the stuff is amazing. Those tooth impressions were then brought home to the laboratory for casting, the end result being epoxy replicas, mounted on a special base and covered by an ultrathin shiny coating of gold/palladium.

We put each replica in the SEM's vacuum chamber and pumped the air out. The microscope worked by focusing electrons into a fine beam and sweeping them across the surface in rows. The beam knocked electrons off a specimen, fewer at lower points and more at higher ones. Those higher points appeared lighter on the screen and lower ones were darker. The result was an image of the surface, assembled point by point, row by row. At first we used Polaroid negatives. We made prints in a darkroom and measured each and every one of the hundreds of scratches and pits on our photomicrographs with a protractor and ruler or measuring calipers. It was a long and arduous process that took hours from start to finish for each specimen. Measurement error was an issue too, especially when scratches were faded and pits overlain on one another. I usually found more pits than Fred Grine, and he recorded more than Mark Teaford. It was actually pretty remarkable that the diet signal was strong enough to rise above the noise at all.

Then there was the 3D-to-2D problem. Just as with a black-and-white photo, shades of gray and shadowing effects give the illusion of depth to an SEM photomicrograph. But just as the subtle details in the Grand Canyon come in and out of view as the Sun passes from east to west across the Arizona sky, depending on your vantage point, fine scratches appear and fade as a specimen is rotated on its stage. You lose information when you represent a 3D surface in 2D, and what you give up depends on angles between the electron beam, surface, and detector. So not only did we get different numbers if we measured a photomicrograph twice, we also got different photomicrographs each time we imaged a surface unless the specimen was positioned exactly the same way.

Microwear Texture Analysis

Then, in 2001, Alan Walker called me out of the blue. He had come to realize that the SEM was not the best instrument for microwear analysis, and told me that it would be my job to find a better one. I couldn't say "no" to my former postdoc mentor, so I started combing the literature. First, I found the work of Chris Brown, an engineer at Worcester Polytechnic Institute in Massachusetts. Brown had developed Kfrax, a suite of computer programs for scale-sensitive fractal analysis. The idea behind it is that the appearance of a surface varies with scale of observation. A coastline might look straight from the vantage point of a satellite, but jagged to someone walking along it. A road might appear smooth to a driver in a car, but rough to an ant trying to cross it. Brown argued that the change in apparent texture with scale is a great measure of surface complexity; I thought it would be perfect for microwear. Recall from chapter 1 that teeth operate on different scales—as guides for chewing at a coarser scale and tools for fracturing at a finer one. Also, Brown had worked on textures of all kinds of surfaces, from chocolate bars to snow skis. Surely someone with such eclectic interests would be willing to give teeth a try. He was.

The search for a new microscope was more frustrating. I felt like Goldilocks trying out beds. One instrument didn't have the resolution we needed. Another couldn't image a large enough area. A third was unable to handle surfaces as curved as teeth. And a fourth wouldn't work with clear epoxy tooth replicas, or even opaque tooth enamel. Eventually I stumbled across the optical confocal profiler. It didn't have all the buttons and knobs of an SEM; it looked more like a microscope you'd find in a high-school biology lab. But there was an important difference. Regular light microscopes illuminate whole fields of view at once, both the bits in focus and those above and below the focal plane. Unless a specimen is completely flat, surfaces appear blurry. Confocal microscopes work differently. They only light the bits in focus. If a specimen is moved up and down in tiny increments, hundreds or even thousands of virtual slices can be made, with only the in-focus parts recorded for each slice. These can be stacked together to make a cloud of points and a 3D model of a microwear surface.

The combination of optical profilometry and scale-sensitive fractal analysis would give us a new way to see and analyze surfaces that matched our understanding of how teeth work. Recall from chapter 1 that most living things try to keep themselves from being eaten by hardening their tissues enough to prevent cracks from starting, or toughening them enough to keep cracks from spreading. If crushing hard foods leaves pits of varying shape and size, surfaces will appear rough at both high and low magnification. Those surfaces would be, in engineering speak, complex. If shearing tough foods leaves scratches aligned in the chewing direction, surface textures will have a preferred orientation. Those surfaces would be, in engineering speak, anisotropic.

I figured that anisotropy and complexity would be great proxies for toughness and hardness of foods, respectively. Chris Brown and his colleagues had developed clever ways of measuring many other aspects of surface texture too. More important, since texture analyses are run on point clouds, we wouldn't have to worry about the 3D-to-2D problem or error noise introduced by measurements of individual features. It would also be a great relief not to have to measure hundreds of tiny scratches and pits by hand each day. My eyes had started to go with age and all the squinting, and I knew my sanity was probably not far behind.

The prospects of this new marriage of dental microwear and surface-texture measurement, or *metrology*, were exciting. We got a couple of grants, bought an instrument, and hired Rob Scott to help work out the methods and scan specimens. Chris Brown and his team began modifying Kfrax for microwear analysis, and Mark Teaford and I headed out to collect dental impressions of primate teeth from collections in museums throughout the United States and Europe. The first step was to confirm that hard-object feeders had higher texture complexity, and those with tough-food diets had more anisotropic surfaces. They did. Leaf-eating primates had, on average, more anisotropic microwear textures, and surfaces of fruit eaters, especially hard-object feeders, were more complex.

But there was more. Remember those sooty mangabeys at Taï and grey-cheeked species at Kibale in chapter 2? They differed in their diets, but both had flat, thick cheek teeth and powerful jaws. We could identify those diet differences much more clearly in the microwear. The Taï mangabey microwear surfaces looked like the surface of the Moon. They were on average more complex, as we'd expect of a hard-object specialist. But the grey-cheeked species had a cluster of lower values, with just a few higher ones. This was a very big deal. It would open the door to understanding not just what our ancestors' teeth were capable of eating, but what they actually ate on a daily basis. It would be a complete shift of focus from adaptation to behavior. That changed how my colleagues and I look at fossil teeth.

The Last Supper Phenomenon and Foraging Strategies

Let's take a closer look at how microwear surfaces form. Primates typically wear through about a hair's breadth, something like 20–200 micrometers (µm), of enamel each year. That may not seem like a lot, but it is when you consider that the microwear scratches we've been discussing are typically less than 1 µm deep. It should come as no surprise, then, that features are wiped away continuously over the life of a tooth as new ones appear. If an animal eats the same sorts of food year-round, the pattern should remain constant. Old scratches and pits should be replaced by more of the same—no big deal. But if diet does change, microwear texture will follow, as new features overwrite old ones. In the end, microwear preserved on a fossil tooth often reflects just the last few meals, a few days, or, at most, weeks before death.

Fred Grine called this the "Last Supper phenomenon." It doesn't bode well for inferring the lifelong diet of an individual, but we can use it to our advantage for reconstructing foraging strategies of a species. Recall that primates pick and choose from the options available to them on the biospheric buffet, and that these can change with the seasons or even the years. If each individual in a sample represents a different moment in time for a member of its species, microwear should give us an idea of variation in diet, given enough specimens. Without the measurement noise of SEM-based microwear studies, we can also be more confident that the variation we see in the microwear signal has real biological meaning.

I remember presenting microwear complexity data for brown capuchins at the ARKUMO (Arkansas, Kansas, Missouri) regional meeting of paleoanthropologists in Lawrence, Kansas, a few years ago. Many of these South American monkeys fall back on hard foods, as do the mangabeys at Kibale. In this case, it's nuts and palm fronds. Their spread of complexity values looks a lot like that of grey-cheeked mangabeys: clustered at the low end, but with a few much higher values. Barth Wright, a specialist on the feeding ecology of capuchin monkeys, was at the meeting too, presenting his data on food fracture properties. The distribution of food hardness values looked just like what we found for the microwear. That sealed the deal in my mind. Microwear *could* help us ferret out details of food choices, foraging strategies, and how a fossil species responded to its changing world.

Hominin Microwear Textures

The first early hominins Rob Scott and I took on were the same ones that Fred Grine had analyzed back in the late 1970s and early 1980s. Differences between the species were obvious, so they would make a good test case for microwear texture analysis. The results were right in line with what Fred Grine had found decades earlier. *Paranthropus* had more complex surface textures overall, and *Australopithecus* had more anisotropic ones. But there was more to the story. There was the spread of values for each species.

Some *Paranthropus* specimens had complex microwear textures, but others looked like *Australopithecus*. Some *Australopithecus* specimens had anisotropic textures, but others looked like *Paranthropus*. The hominins didn't just differ, they overlapped too. I thought back to Ketambe. I thought of the macaques, gibbons, and orangutans all eating the same figs when a big tree was fruiting. But their teeth and jaws were so very different. And while those differences were in line with variation in diet between the species over the long term, all ate fleshy, sugary fruits when there were enough for everyone. Could the same have been true for the South African hominins? They didn't compete directly because *Australopithecus* lived before *Paranthropus*, but the analogy still worked. Perhaps both hominins ate soft, energy-rich foods when they could get them, but each chose items that posed different mechanical challenges when they could not. Maybe this, rather than food preferences, drove the differences in sizes and shapes between *Australopithecus* and *Paranthropus* teeth.

This is a far cry from the highly specialized diet Robinson, or Groves and Napier, or Jolly had envisioned (see chapter 3). *Paranthropus* microwear textures looked like those of brown capuchins. Could nature really have given them big, flat, thick teeth and powerful jaws just for those times that softer, preferred foods were unavailable? Was *Paranthropus* a hominin example of Liem's paradox?[5] When you think of the mangabeys at Kibale, or the gorillas at Bai Hokou for that matter, it makes a lot of sense.

So far we've been talking only about South African hominins. What about eastern Africa? *Australopithecus* and *Paranthropus* were there too, albeit different species. As in South Africa, the eastern African species were separated in time, with *Australopithecus* before *Paranthropus*. They could provide a great independent test of the patterns. We started with *Au. afarensis*, the Lucy species. Its microwear textures looked a lot like those of *Australopithecus* from South Africa—low complexity and a mix of anisotropy values. Lucy wasn't a hard-object feeder either. In fact, eastern African *Australopithecus* had even lower complexity values than did the South African species, which made sense given slightly larger teeth and a thicker skull for the latter.

Results for the eastern African *Paranthropus* sample were, on the other hand, totally unexpected. We considered *P. boisei*, the nutcracker species. Its microwear should have looked like Scott McGraw's sooty mangabeys, or the lunar surface, riddled with deep craters. After all, it had the biggest, flattest, thickest teeth of any hominin. It had incredibly robust jaws and massive chewing muscles. If some South African *Paranthropus* specimens had complex microwear, surely more of the eastern African ones would. But they didn't. None of the specimens I examined had complex, pitted surfaces. Most had simple surfaces covered in wispy scratches. It made no sense. If the nutcracker cracked nuts, there was no trace of it in the microwear.

The Past Is a Foreign Country

We were, in the end, able to get microwear from 30 early hominins recovered along Africa's Great Rift Valley. They all had relatively big, flat teeth and powerful jaws—even *Australopithecus*, albeit not to the degree of *Paranthropus*—but not one had the telltale microwear surface complexity of a hard-food eater. In fact, these hominins had a combination of tooth form and microwear pattern we simply haven't found in living primates, certainly not in mangabeys and capuchins. Perhaps, as L. P. Hartley wrote in the opening sentence of *The Go-Between*, "The past is a foreign country; they do things differently there."[6]

What, then, could possibly have made those surface textures on those teeth? Let's take a step back for a moment and think about how microwear forms. We learned in chapter 1 that hard foods are best crushed between cusp and basin, and tough ones are sheared between upper and lower tooth crests. So tooth shape must also play into the microwear equation. Remember that crushing basins and shearing crests guide tooth movements when opposing teeth are brought together for food fracture. But what if a hominin wanted to eat tough items and didn't have crests to guide its chewing? One option would be to use its teeth like a mortar and pestle, grinding or milling tough food between blunt cusps and opposing basins.

If eastern African early hominins used their molars as mortars and pestles, we might expect chewing surfaces to be covered in scratches running every which way; and that's just what we find. There is some variation in anisotropy, but not the high averages of living folivorous primates, like leaf monkeys and howlers, which have long, sharp crests that constrain and guide tooth movements like scissor blades. Could the nutcracker actually have been a milling machine instead? That could explain the big, heavy jaws; massive chewing muscles; and thick, flat teeth too. It would also explain why its teeth wore so quickly. The wisdom teeth on Mary and Louis Leakey's *"Zinjanthropus"* skull had just erupted when the hominin died; it would have been a young adult. But its other molars were already heavily worn. We don't very often see tooth wear that rapid in primates, certainly not in hard-object feeders. But the steep wear gradient makes sense if *Paranthropus* milled abrasive foods with flat teeth.

Over Hill and Dale

But if other tough-food-eating primates have crests, why didn't eastern African hominins evolve them too? We can think of adaptation as a landscape covered in hills and dales, with elevation representing fitness. The higher you go, the better the adaptation for a given set of conditions (like food properties). Natural selection tends to push species uphill to the nearest peak. There may be a higher one downrange, but you'd have to descend and cross a valley to get there, and that takes a lot of effort. Maybe it requires fewer steps to pump out more enamel and thicken/flatten a tooth than it does to thin it out, raise the cusps, and develop crests. Sure, crests and blades might be more efficient for shearing tough foods, but nature has to work with the materials at hand. So long as *Paranthropus* was better at milling than its ancestors were, it may well have evolved bigger, flatter, and thicker teeth because it's easier to get there from an *Australopithecus* starting point.[7]

The best test for this idea would be to check the microwear of a tough-food-eating, flat-toothed living primate—but there aren't any. There are other mammals we can consider, though. Think of the panda, which eats tough bamboo and has large molar teeth with blunt cusps, at least compared with its closest relatives. The panda has evolved bigger, flatter teeth for a tough-food diet. And its microwear looks a lot like that of the eastern African early hominins: moderate anisotropy and low complexity. Of course, no one would confuse a panda tooth for that of a hominin, so the analogy isn't perfect. Other examples among today's mammals are few and far between. But that's the best we can do with microwear. We could, however, also look for another set of foodprints.

CARBON ISOTOPES

What about chemistry? In chapter 4 we learned that tropical grasses have a higher ratio of the heavy carbon isotope, ^{13}C, to the lighter one, ^{12}C, than do trees and shrubs because of differences in how these plants use the Sun's energy to convert carbon dioxide and water into carbohydrates and oxygen. What about the animal that eats those grasses? Are isotope ratios passed along the food chain such that bones and teeth get their proportions of heavy to light carbon from the plants used as raw materials to make them? Do these proportions remain unchanged over deep time so fossils can preserve the telltale chemical signatures of diet? Researchers have built careers on addressing these questions and, in the process, have developed a whole new way to reconstruct the diets of animals alive in the past.

Relating Isotopes to Diet

Nikolaas van der Merwe, a South African–born archaeologist and physicist, was among the first. Van der Merwe didn't start out interested in using carbon isotopes to reconstruct diet. In fact, back in the early 1960s when he was in graduate school, botanists had barely worked out the C_3 and C_4 photosynthetic pathways that today allow us to use carbon isotope ratios to separate trees, shrubs, and herbs from tropical grasses and sedges. Researchers at the time were using carbon isotopes in the process of dating archaeological remains,[8] not for figuring out what people or animals ate in the past. That's how van der Merwe at first used them. His doctoral dissertation at Yale was on the use of carbon isotopes to date ancient iron artifacts. But the skills he acquired in New Haven would serve him well as his research expanded and took him in additional directions over the years to come.

Van der Merwe took his first teaching job at Binghamton University in upstate New York in the late 1960s. Archaeologists working there at the time were focused on the origins of agriculture in the region, when prehistoric peoples began planting and eating maize. But it was difficult to get good dates from maize kernels at archaeological sites; they consistently came out a couple of hundred years younger than wood charcoal in the same deposits.

This was also about the time that botanists worked out the differences between C_3 and C_4 photosynthesis, and when they realized that C_4 tropical grasses have higher ratios of heavy to light carbon atoms than do C_3 trees or bushes. This, in fact, explains why van der Merwe's old corn kernels seemed to be too young. It turns out that during chemical reactions, plants discriminate against heavier carbon. But C_4 plants, which include maize, discriminate less than C_3 plants do against both ^{13}C and ^{14}C. That's why maize ends up with relatively more ^{14}C in its tissues to begin with, and why the ratio of unstable (^{14}C) to stable carbon is higher in maize than in wood after a given period of decay.

This got van der Merwe to thinking. If early farmers in upstate New York ate mostly maize, they should have more ^{13}C and less ^{12}C in their bones than the hunter-gatherers that came before them. After all, wild plants that far north are all C_3, with relatively fewer ^{13}C atoms. To check, van der Merwe teamed up with John Vogel, an expert in radiocarbon dating at the National Physical Research Laboratory in Pretoria. The two ran samples from prehistoric human rib bones shortly after van der Merwe returned to his native South Africa (to take a professorship at the University of Cape Town) in the mid-1970s. The difference in stable carbon isotope ratios between the hunter-gatherers and maize farmers was just as predicted. Van der Merwe and Vogel had developed an entirely new tool for inferring diets of past peoples.

As a sidenote, van der Merwe and Vogel were not alone in recognizing the value of carbon isotopes for diet inference. At about the same time, Michael DeNiro and Sam Epstein at Caltech collected stable carbon isotope data for Henrick Hoeck's Serengeti hyraxes—the same specimens that Alan Walker was studying for microwear. DeNiro and Epstein's results showed the grass-eating rock hyraxes to have relatively more ^{13}C than the bush hyraxes that lived alongside them. So not only could microwear distinguish between tropical-grass-grazing and tree- or bush-browsing animals, so too could carbon isotope analyses. The two studies were published back to back in the journal *Science*.[9]

Relating Isotopes to Hominin Diets

Back in South Africa, van der Merwe continued his work on isotopes and diet. He set up his own laboratory and built a research group in Cape Town. One of his students, Julia Lee-Thorp, was especially interested in using stable isotopes to infer the diets of fossil species. It was unclear at that point whether it would even work. A lot can happen between death and the time a bone or tooth is found by a paleontologist. Buried bones and teeth can change dramatically with soil moisture, pH, heat, and exposure to microbes and inorganic compounds. What effect would these have on isotope ratios? DeNiro and Margaret Schoeninger, then at UCLA, looked at ratios of ^{13}C to ^{12}C in ancient human bone mineral from Mexico's Tehuacán Valley, for example, and they were all over the map. They argued that carbon from the atmosphere or the ground must have made its way into the bones, and that led some researchers to question the reliability of carbon isotopes for diet work.

But Lee-Thorp and van der Merwe pressed on. After all, stable carbon isotope analysis had separated maize farmers from their hunter-gatherer predecessors in upstate New York. They worked to develop a technique that would remove contaminants, and tried samples of other species. This time, it was bones dating back as far as three million years ago: fossil giraffes and antelopes whose living relatives today are pure browsers. While the carbon isotope ratios were not exactly as would be expected for browsers, no one would confuse those fossils with grazers. The pretreatment had helped a lot. Even better, Lee-Thorp and van der

Merwe were able to eliminate the effects of chemical changes almost entirely by switching from bone to tooth enamel. By the late 1980s, it was clear that stable isotope ratios of carbon in tooth enamel could remain nearly unchanged for a very long time. Because the carbon used to make teeth comes ultimately from food eaten during dental development, those ratios could tell us something about the diets of individuals that lived millions of years ago.

Demonstrating the potential of the technique was key to taking the next big step—analyzing an early hominin. Our fossil ancestors are rare and priceless jewels, a legacy to all humanity. To measure isotopes you have to pry or drill a flake of enamel from the crown, grind it into powder, dowse it with acid, and measure the carbon isotopes in a mass spectrometer. Back in the 1980s, it took a large chunk, enough to cover half a tooth cusp. Who knows what evidence about the lives of early hominins might be lost to future generations by destroying even one precious sliver of enamel? Surely it would not be worth the risk unless it was clear that valuable information would follow. Now it seemed clear.

Lee-Thorp began with *Paranthropus* from Swartkrans (see chapter 3). Grine's original microwear studies had suggested that these hominins ate hard plant parts like nuts, or perhaps the underground bits of root vegetables. Those are C_3 plants with a lower ratio of ^{13}C to ^{12}C, and the hominins that ate them should have had a correspondingly lower ratio in their enamel. But the numbers didn't come out quite as expected. *Paranthropus* had too much ^{13}C relative to ^{12}C in its enamel for it to have had a C_3 diet. It wasn't contamination, either. Kudu and baboons from the same site had less heavy carbon, just like their living relatives. But they weren't grazers either. In fact, the proportions of ^{13}C to ^{12}C atoms suggested the hominins got about 25% of their carbon from C_4 plants. Did *Paranthropus* supplement a predominantly C_3 plant diet with the grasses spreading across the veld at the time? Lee-Thorp thought of another possibility. There's another way to get a C_4 signal—by eating animals that eat grass. After all, carbon passes from plant to herbivore to carnivore, so eating a grazer should leave about the same amount of heavy carbon in teeth as eating grass. Perhaps, then, *Paranthropus* included meat in its diet. It did, after all, have a large brain to fuel, at least compared with the average ape, and grass

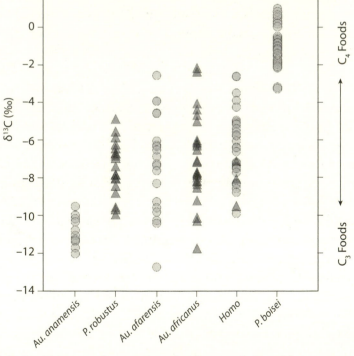

5.3. Julia Lee-Thorp sampling an *Australopithecus* tooth for stable isotope analysis (*top*) and carbon isotope data for Plio-Pleistocene hominins (*bottom*). Eastern African hominin specimens are represented by gray circles, and South African ones by black triangles. Top image courtesy of Julia Lee-Thorp, and bottom one modified with permission of Matt Sponheimer (see Matt Sponheimer, et al., "Isotopic Evidence of Early Hominin Diets," *Proceedings of the National Academy of Sciences* 110, no. 26 (2013): 10513–18, for details).

isn't a very good source of ready energy for a primate. Meat is much more easily digested.

Over the next few years, more and more South African hominins were added to the sample. Lee-Thorp teamed up with Matt Sponheimer, then a graduate student at Rutgers University, to study carbon isotopes of *Australopithecus* teeth from Makapansgat. And van der Merwe got more samples of *Australopithecus* from Sterkfontein. In the end they had about 50 specimens, mostly *Australopithecus* and *Paranthropus*, but also a few early *Homo* (we'll discuss those in chapter 6). Tooth after tooth, the story was the same—a diet dominated by C_3 plants, but with a hearty helping of C_4-derived carbon on the side. A couple of things became clear. First, the South African early hominins were not eating like chimpanzees. They ate plenty of tree and bush parts, but also grasses, or perhaps animals that ate those grasses. Second, *Australopithecus* and *Paranthropus* teeth had similar mixes of carbon isotopes from C_3 and C_4 plants. If their diets differed, it wasn't in a way that stable carbon isotopes could detect.

Eastern African Samples

But what about eastern African hominins? The custodians of fossil hominin vaults in Dar es Salaam, Nairobi, and Addis Ababa were more wary of destructive sampling, and it took a long time to convince them to open the doors for isotope study. But techniques developed by Lee-Thorp in the 1990s meant that little more than a sugar grain's worth of enamel would be needed, and sampling could be limited to broken bits of tooth. The curators finally relented in the late 2000s.

The first eastern African hominin isotope data came from Tanzania, published in 2008 by van der Merwe and colleagues.[10] It was on only five specimens—two *Paranthropus* and three early *Homo*—but the results were surprising. The *Paranthropus* specimens had a much higher ratio of ^{13}C to ^{12}C, presumably because of more C_4 plants in their diet, than was typical of South African hominins. Two specimens weren't much to go on, but the results were tantalizing—especially since they looked different from the early *Homo* values, which matched those from South Africa. The following year, Tim White and his colleagues published carbon isotope values for *Ardipithecus*, a very early hominin from Ethiopia dated to about 4.4 mya.[11] *Ardipithecus* was also different from the South

African specimens, but in the opposite direction. It lacked evidence for C_4 plants in its diet. Perhaps it lived before hominins ventured from the forests or woodlands into more open grasslands.

All of a sudden, early hominin isotope values weren't all the same any more. Researchers could separate species by carbon isotope values, and start to think about what those differences meant. But more numbers would be needed to confirm these results, and there were a lot more species to analyze in eastern Africa. Matt Sponheimer, now on the faculty at the University of Colorado, and Thure Cerling (we met him in chapter 4) led the charge, beginning with *Paranthropus*. Were van der Merwe's numbers typical of the species, or were they just a couple of weird outliers? Sponheimer and Cerling sampled about two dozen teeth from Kenya to check, and the results were clear. *Paranthropus* from eastern Africa *was* different. For that species, grasses or sedges were not a supplement; they made up at least three-quarters of the diet.

As more eastern African species were added to the sample, a pattern began to emerge. There were three groups. *Ardipithecus* and the earliest of the *Australopithecus* species (*Au. anamensis*, dated to 4.2–3.9 mya) had a lower ratio of ^{13}C to ^{12}C. They evidently mostly ate parts of trees and bushes, like chimpanzees do today. The slightly later *Au. afarensis* had a mixed diet, like the early hominins from South Africa. And as we've already mentioned, the even later nutcracker species, *Paranthropus boisei*, had the telltale signal of a grass or sedge eater. Perhaps, then, hominins began to eat more savanna foods shortly after 4 mya, and grasses, or sedges, became an increasingly larger share of the diet over time, culminating in *Paranthropus*. It makes for a compelling story. Of course, it's probably not quite this straightforward, at least not in South Africa, because *Australopithecus* and *Paranthropus* fossils there have similar average carbon isotope ratios.

BRINGING IT ALL TOGETHER

So now we have isotopes and microwear for *Australopithecus* and *Paranthropus*, and we have them for species from both South Africa and the Great Rift Valley. We have their fossil remains, and a basic understanding of relationships between tooth form and function, or how teeth work. We know something about the role of teeth in food choice, and how food availability in different habitats can affect that role. And we've seen the ebb and flow of climate change over deep time, and how that changed the worlds of the early hominins and the options available on their biospheric buffets. Let's lay out the pieces on the table and see how we can fit them together.

The first thing that jumps out is how much more complex the story is than Robinson, Tobias, Jolly, or Groves and Napier imagined it to be back in the day. As we add pieces to the puzzle, more and more seem to be missing. Science works that way sometimes. The great eighteenth-century philosopher Immanuel Kant wrote, "Every answer given on principle of experience begets a fresh question, which likewise requires its answer."[12] Foodprints give us a new vantage point, but when we reach the crest of the hill we see a higher peak to climb, off in the distance. The past becomes increasingly muddled as new facts emerge. But that's good. It means we're probably on the right track. The Earth system today, especially the biosphere part, makes for a pretty tangled web of relationships and interactions. The worlds of our ancestors must have been much the same, and the system becomes even more complex when we add a time element to the mix.

The story is complicated in part because the various elements each represent different aspects of diet sampled at different time scales. Isotopes tell us whether an early hominin got its carbon from C_3 plants, C_4 plants, or some combination of the two while its teeth were forming. That carbon could have come from plants, prey animals that ate them, or both. Microwear textures may tell us whether a hominin ate hard foods or tough ones in the days or weeks before death. Tooth size, shape, and structure, on the other hand, offer up information about potential. Could a hominin have eaten especially hard, tough, or abrasive foods? These different lines of evidence needn't, and indeed shouldn't, give us the same details. But we need them all to see the big picture.

All of the species we've considered had big, flat, thick teeth; strong jaws; and well-developed chewing muscles, though to different degrees. *Paranthropus* took these things to an extreme compared with its *Australopithecus* predecessors. Conventional wisdom would have us believe that changes in the sizes and shapes of teeth and jaws mean increasing specialization for hard, brittle foods from *Australopithecus* to *Paranthropus* at about the time grasslands spread though eastern and southern Africa. Of course, it wasn't that simple. But this basic story line gives us a place to start and some context with which to interpret our confusing and seemingly inconsistent foodprint signals.

Species of early hominin of the same genera in eastern and southern Africa differ in at least some cases in both microwear and isotopes. *Australopithecus* and *Paranthropus* from eastern Africa differ in isotopes but not microwear. And those in South Africa differ in microwear but not isotopes. So it looks as though in South Africa the C_3 and C_4 food mix remained steady over time, but there was increasing consumption of hard objects, at least as fallback foods. This makes sense when we compare the teeth and jaws of these hominins. For eastern African hominins, though, the foodprints suggest increasing consumption of grassland foods. This also makes sense when we compare the teeth and jaws, albeit in a different way, if grasses or sedges require heavier grinding. Grinding therefore seems to have been the modus operandi for the Plio-Pleistocene hominins of eastern Africa—especially for the robust species, judging from its scratched up, quickly worn teeth.

It is as if nature found the same solution to different problems. The more specialized teeth and jaws of *Paranthropus* would have served them well for breaking mechanically challenging foods wherever they lived, but in the south it was hard objects, and in the east it was tough ones. This is something that my colleagues and I had never expected, or even considered as a possibility. But it makes sense when you think in terms of adaptive landscapes, and the place you're starting from. Remember that nature is limited to the raw materials at hand, and there's only so much you can do with them.

So what about the lack of change from *Australopithecus* to *Paranthropus* in microwear in the east or in isotopes in the south? How can we reconcile that with changing anatomy? The answer may lie in a well-known concept in evolutionary biology called the Red Queen hypothesis. As the Red Queen told Alice in Lewis Carroll's *Through the Looking Glass*, "Now, here, you see, it takes all the running you can do, to keep in the same place."[13] There are many cases of species becoming better over time at how they earn a living, under pressure of competitors, predators, or prey doing the same. Richard Dawkins is fond of the rabbit and the fox example from Aesop's fables. It's the evolutionary arms race we discussed back in chapter 1. Fine-tuning anatomy for heavy chewing need not mean a change in diet. Sometimes species change their teeth and jaws just so they can do a better job of eating whatever it is they eat.

Consider Basil Cooke's fossil pigs. Recall from chapter 3 that their third molars got longer over time, reliably enough for Cooke to use them to date fossil deposits. The idea was that pigs evolved larger teeth for chewing more low-quality, abrasive grasses as savannas spread across the eastern and southern parts of the African continent. In fact, Thure Cerling and his colleague John Harris found that for at least one pig lineage, *Sivachoerus-Notochoerus* for those keeping track, longer teeth actually had higher proportions of ^{13}C, so later species indeed ate more C_4 plants.[14] This is essentially the pattern we see for the early hominins in eastern Africa, at least as far as the isotopes go.

On the other hand, the pig lineages *Metridiochoerus* and *Kolpochoerus* show a different pattern. They started out with a C_4 diet in the first place, and stayed that way. But their teeth *still* got longer over time, and presumably better adapted to the tough, abrasive grasses that members of the lineage had been eating all along. So, in one group increasingly specialized anatomy was matched by a change in diet, but in the other it wasn't. Perhaps in eastern Africa, then, hominins ate tough foods straight through the Plio-Pleistocene, but in the south they increased their hard-object consumption. In contrast, in the east they shifted from C_3 to C_4 plants, but in the south they kept a steady mix of the two. In both cases, larger, thicker, stronger teeth and jaws would have done a better job of breaking fracture-resistant foods, whether they were hard or tough, and whether that meant a shift in diet or just more of the same.

What we're left with, then, looks kind of like an evolutionary "free-for-all", where some hominin species have different teeth and jaws for similar diets and others have similar teeth and jaws for different ones. None of us had seen these results coming, though we probably would have had we been thinking like ecologists. Consider the living primates. When we compare red-tailed guenons and mangabeys at Kibale, or mangabeys at Kibale with those in the Taï forest, it all starts to make sense. Recall again that the two species at Kibale eat the same foods most of the time despite having different teeth and jaws. And the two mangabeys at the different sites have very different diets most of the time, despite having similar teeth and jaws. In these cases, the sizes and shapes of teeth and jaws are not enough—we've got to watch actual individuals eat to know how they use their teeth. Of course, we can't watch fossil hominins eat, but that's where foodprints come in. They are, as was suggested at beginning of this chapter, our portholes to the past.

Our Changing World

Matching hominin tooth form and foodprints can help us understand what individuals in the past ate, but not necessarily why they ate it. That's where concepts from evolutionary theory—like Liem's paradox, adaptive landscapes, and the Red Queen hypothesis—come in. We've learned that specialized teeth can reflect a specialized diet or a more generalized one. This is because natural selection seems to focus more on foods that pose the greatest challenge and less on their proportion in the diet. We've learned that teeth can evolve for new foods, or for more of the same, if a change in crown shape increases the efficiency with which they fracture whatever it is they have to break. We've learned that a given tooth form can evolve for different functions, so long as it does a better job than its predecessor at each of those functions. The living mammals give us many examples. There may be examples of each in the early hominins too. At least, that's how I read the foodprints.

So how does all this square with climate change over time in Africa, and the habitat proxies of deMenocal, Cerling, Potts, and Maslin introduced in chapter 4? Recall that there was much more to environmental change over the course of human evolution than a simple shift from warmer and wetter conditions to cooler and drier ones. We've learned that climate varies according to a cyclic pattern that follows the ever-changing relationship between Earth and Sun. That translates to environment differently in different places—depending on our restless Earth, especially how plate tectonics affects regional and local landscapes. Yes, Africa trended cooler and drier over time, but there were also increasingly intense swings in climate and habitat type.

The savanna hypothesis, as Dart originally envisioned it at least, simply isn't big enough to accommodate the combinations of early hominin tooth form and foodprints considered here. But the idea of climate swings back and forth, in different places at different times, squares much better with the lack of a single explanation for the trend toward bigger, flatter, thicker teeth and more powerful jaws. If those teeth could lead to a more generalized diet or a more specialized one, to more tough foods or more hard items, in a closed forest setting or a more open grassland, it only makes sense that *Paranthropus* would have evolved, not as a specialized nutcracker, but as whatever it needed to be where and when it lived. Its teeth and jaws let it cover all the bases. And the nutcracker actually did quite well for itself. It survived and thrived in its increasingly unpredictable and challenging world for more than a million years. We should be so lucky.

But that's only part of the story. The earliest members of our own genus, *Homo*, come next. Those hominins found an entirely different solution.

CHAPTER 6

What Made Us Human

Lake Eyasi is set in an idyllic spot on the floor of the Great Rift Valley, less than 20 miles from Olduvai Gorge as the crow flies. The Hadza people who live around the lake today are the last foragers in Africa to make a living almost entirely on wild foods. They have a wonderful creation story. Ishoko, the Sun, divided a troop of baboons into two groups. She sent one to collect water and the other to gather food to bring back and share. The food gatherers returned after a time with their bounty, but the water collectors did not. Ishoko went looking and found them drinking and playing in a distant stream. She rewarded those who had brought back food by turning them into the Hadza people, and the others were fated to become their favored prey.

What made us human? The answer to this question is the Holy Grail of paleoanthropology, and researchers have been on the quest a very long time. Many consider art, language, self-awareness, or empathy when they think about the differences between us and other animals, but none of these things came up when I asked the Hadza men and women of Sengali Camp, up on Gideru Ridge. They are part of the larger community of life that surrounds them, and they feast or famine today directly from the biospheric buffet, just as they have for countless generations. Most of the rest of us are too far removed to recognize that an important part of what made us human was how our forebears chose to fill their plates. How could we realize it, when our meat comes wrapped in cellophane, and our vegetables are vacuum-packed in aluminum cans?

Paleoanthropologists have thought a lot about this. A common theme involves a fundamental change in how our ancestors earned a living in response to environmental triggers around the onset of the Pleistocene (see chapter 4). Hunting made us human. Gathering made us human. New tools to collect and process food made us human. Cooking made us human. In all of these cases, diet is central to the tale. In this chapter, we consider ideas about what made us human from the perspective we've built on teeth, diet, and a changing world. But there's more of interest here than the Grail itself, there's also the story of the quest, and the scientists leading it. Let's head north from Hadzaland back into Olduvai Gorge and have a look.

IN SEARCH OF THE EARLIEST HUMANS

Olduvai Gorge is carved into a broad basin on the eastern edge of the Serengeti. The main ravine runs nearly 30 miles in length, from Lake Ndutu eastward to the rising slope of the Ngorongoro Highlands. It's not quite a canyon, but it is steep sided in places, and it cuts through a total of about 300 feet—two million years of sediment deposited as ancient lakes and streams waxed and waned through the distant past. Today it's hot and arid much of the time, with little to distract beyond the occasional bell on a Maasai goat or cow passing through. But to paleo-anthropologists it's a magical place. The floor of the gorge is riddled with small gullies, and animal fossils and stone tools erode out everywhere.

Louis Leakey began collecting bones and stones at Olduvai in 1931 but, except for a few scraps, the toolmakers themselves alluded him. Then his wife, Mary, found the nutcracker man skull in 1959. As their second son, Richard, recalls, "He knew there were tools, and he knew that those tools had been made by somebody, so when somebody was found, he assumed that it had made the tools."[1]

Jonny's Child

The following year Richard's older brother Jonathan made another important discovery. It was only a couple of hundred yards away from the spot where Mary had found that first skull. Jonathan had gone off to look for fossils on his own and stumbled upon a saber-toothed cat jaw eroding from a nearby gully. The Leakey team sieved through the surface debris around the find and dug a trench in search of more of the cat, but instead struck paleoanthropological gold. Jonathan unearthed a couple of large chunks of a second hominin skull, a lower jaw with teeth, and several hand bones. Louis and Mary called the new specimen

"Jonny's child." Even though only the sides of the cranium were preserved, the new braincase was clearly larger than that of the nutcracker man. The jaw was slender, and the teeth were too small and narrow to match those of Mary's first skull. This was a new species, a more humanlike one. Just as John Robinson had found at Swartkrans more than a decade earlier (see chapter 3),[2] there were two hominins eroding from the early Pleistocene deposits at Olduvai—one with a bigger jaw and teeth and the other with smaller ones, closer in size and shape to ours.

Over the course of the next three years, the Leakey family unearthed many more hominin fossils at Olduvai. Louis asked Phillip Tobias, who was working on the nutcracker man, to describe and analyze the other skulls and teeth too. He gave the hand to John Napier, who had worked on hand bones of a fossil primate the Leakey crew had discovered a decade earlier on Lake Victoria's Rusinga Island. In 1964, the three described and named the new species. Its brain was twice the size of a chimpanzee's and 25% larger than the nutcracker man's. Its teeth were small and narrow, and its hand bones were very humanlike. Grasp a pencil and notice how the tips of your thumb and other fingers face one another. That's how Jonny's child would have held one. A chimpanzee holds a pencil like we hold car keys. If you try writing like that, you'll understand the difference in precision. Now Louis was sure the new species—with its bigger brain, smaller teeth, and dexterous fingers—rather than the nutcracker man, had made the tools at Olduvai. To underscore this, he and his colleagues named the new species *Homo habilis*, taken from the Latin meaning able, handy, mentally skillful, and vigorous.[3]

Defining a Genus

It was a big deal to name a new species. By the early 1960s, John Robinson had realized that the more humanlike hominin at Swartkrans he and Robert Broom had named "*Telanthropus capensis*" (see chapter 3) was better included in *Homo erectus*.[4] To him it seemed that *Australopithecus* graded directly into *H. erectus*. He believed there just wasn't enough room between the two for another species.

But it was an even bigger deal to include the new species in *Homo*. That required Leakey and his colleagues to redefine the human genus. At the time, most researchers conceded that the species *erectus* should

A   B

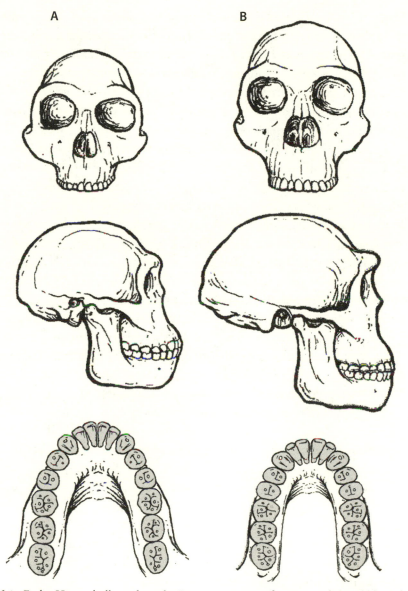

6.1. Early *Homo* skulls and teeth. Reconstructions of crania and mandibles of A. *Homo habilis*, and B. *Homo erectus*. Images courtesy of John Fleagle.

be included in *Homo*.[5] After all, its brain was about 1000 cm³, twice the size of that of *Australopithecus* or *Paranthropus*, and three-quarters that of yours or mine.[6] But *H. habilis*? The *H. habilis* brain was thought to be less than half the volume of ours.[7] To Leakey and his colleagues, though, its small teeth, dexterous hands, and stone tools pointed to a new, more humanlike way of life. To them, it had earned its place in the human genus, big brain or not.

But what does it mean to be part of the human genus? To answer that question, we need to step back another decade. The leading evolutionary biologist of the day, Ernst Mayr, addressed the issue directly at the Cold Spring Harbor Symposium on Quantitative Biology in 1950. In those days it seemed that every time someone found a new hominin it was given a new name. There were more than a dozen genera and many more species being bandied about in the literature, with little rhyme or reason to how they were named in many cases. Mayr said, "Enough!" He argued that all the hominins known at the time should be sunk into one genus, *Homo*, and three species: "*H. transvaalensis*" (which included *Australopithecus* and *Paranthropus* from South Africa), *H. erectus*, and *H. sapiens*. Few others at the time were willing to go that far, but Mayr did bring up an important point that was taken to heart. A genus has to mean something.

Mayr believed in defining genera as groups of closely related species, but he also thought that species in genera had to share fundamental aspects of their way of life that set them apart from others. They had to share their perch on a distinctive adaptive plateau, a unique adaptive zone if you will. For Mayr, upright posture and walking on two legs defined that zone. For many others, though, it was a big brain and the intelligence that resulted from it. But for Leakey, Tobias, and Napier, *H. habilis* had scaled the cliff with stone tools. This ushered in a new way of life, and a new relationship with the world around it. The nimble hands and small teeth were proof.

But How Many Species?

Field paleontologists have continued to scour early Pleistocene deposits in the years since the Leakey team first unearthed Jonny's child at Olduvai. They've uncovered dozens of specimens in the past half century that most paleoanthropologists today agree are neither *Australopithecus* nor *Homo erectus* but, rather, something in between. Robinson was clearly wrong about the lack of space between *Australopithecus* and *H. erectus*. In fact, there are probably at least two species squeezed in there. But how can we know? Fossils don't come out of the ground with name tags on them. We lay them out on a table and compare them to others that were found before. Are two teeth similar enough to be included in the same species, or genus for that matter? There are some rules that most agree on, but researchers interpret those rules in different ways, and there are often judgment calls to be made. It can be a messy business because no two individuals are the same, and one species can seem to blend into the next.

Where do we draw the line? Mayr believed there couldn't be more than one hominin at a time, but now the evidence to the contrary is too overwhelming to ignore. How many species of early *Homo* were there? My colleague here at Arkansas, Mike Plavcan, once measured the teeth of a bunch of guenons to see whether he could separate them by species. These are closely related monkeys, several of which live together in the forests of sub-Saharan Africa. They're all about the same size, but they are genetically distinct, and clearly look, sound, and act like different species. His experiment was kind of like throwing a bunch of teeth in a big paper bag, shaking it up, and spilling them out on the table to see how well they could be sorted. In the end, Plavcan couldn't do it. There was just too much overlap. And if we can't sort out living monkeys by basic tooth measurements, we need to wonder about our confidence in doing it for closely related fossil hominins.

Fortunately, though, there's more to teeth than crown measurements, and they can often be separated by pattern of bumps and grooves. It helps when teeth come out of the ground attached to skulls too. Also, some species are easier to distinguish than others. We do the best we can. But Plavcan's exercise explains why paleoanthropologists sometimes disagree on the number of species represented by a pile of fossil teeth, and which tooth belongs to which. Nevertheless, many of us recognize two species between *Australopithecus* and *H. erectus*—*H. habilis* (about 2.4–1.4 mya) and *H. rudolfensis* (perhaps 1.9–1.8 mya).[8] While *H. habilis* appeared before *H. erectus* (which ranged between about 1.89 mya and 143 kya), the two overlapped for at least half a million years. It's hard to know where *H. rudolfensis* fits into the picture because there aren't many specimens; the species seems to be similar to *H. habilis*, yet still distinct from it.[9]

THE TRANSITION TO *HOMO*

We've now got our cast of characters, with early *Homo* likely in place by the beginning of the Pleistocene if not before. We have early stone tools, which are seen by many as evidence of a new, burgeoning adaptive zone. But how can we use all this to begin to explore what made us human? Archaeologists have developed a whole discipline around teasing past behaviors from lithic artifacts and other bits of durable trash our messy ancestors left behind. And while monkeys and apes may not be great models for *Homo* behavior because they don't make stone tools, we can look to humans, at least to those of us who still hunt and gather wild foods, like the Hadza. There are many things that separate our adaptive zone from those of the other apes. An important one that researchers noticed from the very onset was that human foragers tend to eat more and larger prey animals.

The Scatter and the Patches

Meat was thought to be important to the story of our genus well before *Homo habilis* was first discovered at Olduvai Gorge. Recall from chapter 3 that John Robinson made it central to the narrative after he found *"Telanthropus."* Longer dry seasons led our ancestors to hunting, tool making, and larger brains, all of which carried them past the *Australopithecus* stage of evolution. From the outset Louis and Mary Leakey considered the broken bones strewn across the "living floors" at Olduvai to be the remains of prey animals hunted by the hominins, just as Raymond Dart had assumed for the sites in South Africa. But times were changing. Bob Brain was beginning to look much more closely at bones recovered from the South African sites, and it seemed that those hominins, at least, were more likely the hunted than the hunters (see chapter 3).

Olduvai was different; but still, the hand bones of Jonathan's hominin were not found clutching stone tools, and its teeth were not biting into animal bones either. Everything was a jumble of scattered bits eroding out of the ground or dug from beneath the surface. It was easy to imagine that hominins fashioned the tools and hunted the animals, but to make any real sense of the evidence would require a lot more work. Surely the scatter and patches of teeth, bone, and tools at sites like Olduvai could provide important clues about the lives of our ancestors if only someone could figure out how to read them.

South African–born archaeologist Glynn Isaac led the charge. Louis Leakey had appointed him warden of Kenya's prehistoric sites in 1961, just months after Isaac received his BA from Cambridge.[10] This began Isaac's long association with the Leakey family, which would last until his untimely death in 1985. He took a job at Berkeley in 1966, but returned to Kenya regularly, mostly to codirect research at Koobi Fora, on the eastern edge of Lake Turkana, with Louis and Mary's son, Richard. It was there that Glynn Isaac sought evidence for and honed his *home-base hypothesis*.

He started by assembling a list of differences between humans and other primates. We walk on two legs and carry tools, food, and other things from place to place. We communicate with language to broker social relationships and exchange information about the past and future. We share and trade food as part of our "corporate responsibility." We, at least those among us who still forage for wild foods, devote more time to hunting, and take large prey, sometimes even larger than us. Finally, we maintain a home base, or central place, where we bring and divvy up the bounty collected while hunting and gathering. We share meals. Other primates don't even have meals. They tend to eat food when and where they find it, and at most tolerate some scrounging.

Had these differences emerged by the time early *Homo* walked the landscape at Olduvai and Koobi Fora? If so, could this be read in the scatter and patches of fossils and tools? Isaac thought so. Some sites looked like quarries, with lots of stone flakes but little bone. Others had bones of a single large animal or just a few, and many chipped stones. Yet others had hundreds or even thousands of stone tools and bones representing many different animals. Here was his evidence for human-like behavior nearly two million years ago. These were the factories, kill sites, and home bases to which hominins returned, food in hand. It seemed that the way hominins used the landscape was organized around the transport and sharing of meat.

Isaac committed the rest of his career to testing his home-base hypothesis and working out the details. He recruited a crackerjack team of graduate students at Berkeley in the early 1970s and began to divvy up tasks. One would sort the stone tools and another would figure out how they were made. One would refit flakes found in different areas to track hominin movement across the landscape, and another would work out whether any had been spread by natural agents, like streams or rivers.

He gave the animal bones to Henry Bunn. Isaac first met Bunn at a conference in Nairobi in 1973. Bunn was an undergraduate geology major at Princeton and had been prospecting deposits around Lake Turkana with Vincent Maglio.[11] He was inspired by Isaac, and quickly realized his future was to be at Berkeley. It was a natural fit because his training was in paleontology and how fossil assemblages were formed. It would be his job to impose order on the seemingly chaotic jumbles of teeth and bones to prove that the animals found with the hominins were indeed their prey.

But where to start? Bunn began once he got to Berkeley by using stone flakes to cut and break up cow parts from meat departments at local grocery stores. He also combed through bones from much later sites that were clearly food remains. Bunn found that stone tools used in butchery left cut marks in telltale places, where muscles attached and ligaments connected fleshy limbs. Those marks were clearly different from gnaw marks left by rodents or large carnivores, and from scars or cracks from trampling, weathering, or damage after burial. Hammerstones used to break open bones for access to the nutritious marrow inside also left distinctive fracture patterns. If hominins were processing animals for meat and marrow, he would be able to see it. He had done his homework and was ready to start.

Isaac arranged for Bunn to study the animal remains from Mary Leakey's excavation at the original nutcracker man site. There were thousands of stone tools and bones of all sorts of species from the original excavations. It was the perfect place to start. Bunn set up shop in a small office at the National Museum in Nairobi, and began gathering trays filled with bits of fossil bone from the collections. He found his first cut mark within five minutes and, in doing so, definitively connected the animals, the stone tools, and the hominins at Olduvai for the first time.

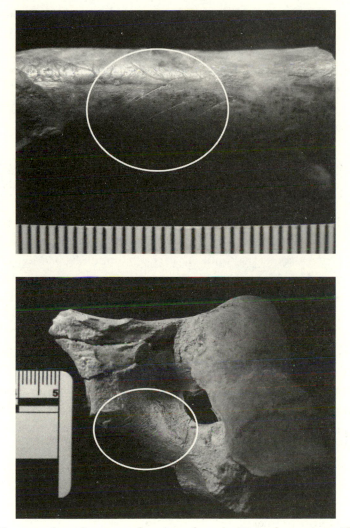

6.2. Upper-arm-bone fragments of antelopes from the *Zinjanthropus* site at Olduvai Gorge. Note the cut marks indicated by the circles. Images courtesy of Henry Bunn.

The evidence was indisputable. The hominins *had* used the stone tools to butcher the animals. But could the bones help him evaluate Isaac's home-base hypothesis?

He dug deeper. Bone after bone, tray after tray, day after day, Bunn sorted through the thousands of broken bits. It was painstaking and tedious work. These were not precious hominin fossils, each carefully identified, described, and cradled in its own foam-lined tray in the museum's bombproof vault. Many were unidentified fragments, still covered in dirt from the original excavations, and bundled together by the hundreds in large plastic bags, grouped by the trench Mary Leakey and her team had dug them from more than a decade before. Bunn had to piece these back together to identify individual animals and figure out how many bore cut marks and hammerstone fractures, as well as where they were located on each bone.

In the end, he found hundreds of cut marks and hammerstone fractures on the bones of dozens of animals from the deposits at Olduvai and also Koobi Fora. The hominins had a penchant for antelope meat and marrow, though they ate many other types of animal too, such as pigs, hippos, horses, giraffes, and even elephants. There could be little doubt that by about two million years ago, hominins had joined the guild of the large carnivores.[12] They had come a long way from the frugivorous apes that had inhabited primeval forests of earlier epochs.

The cut marks were concentrated mostly on the fleshy parts, but there were some at the ends of limb bones, like the elbow and knee, where ligaments would have held them in place. And there were lots of arm and leg bones at the sites Isaac called home bases. There were whole limbs for the smaller animals, but mostly meat-bearing upper parts for the larger ones. It was as if the hominins had killed and butchered their prey elsewhere and schlepped the parts they could carry back home. Bunn's evidence looked great for Isaac's home-base hypothesis.

But the bones were also riddled with carnivore tooth marks, and some researchers suspected that large cats, dogs, and hyenas, rather than hominins, had killed the animals and collected their bones. There were all those hammerstone fractures too. Maybe the hominins weren't actually hunting, but instead scavenging, harvesting marrow from the bones left behind after carnivores had eaten the flesh and other soft tissues of their prey. Rick Potts (see chapter 4) and his colleague Pat Shipman, of Johns Hopkins University at the time, took a closer look using a scanning electron microscope. Some bones had marks from both stone tools and carnivore teeth, and these even overlapped in a few—both cut mark over tooth mark and tooth mark over cut mark. It seemed almost like a hunting and scavenging free-for-all, with both hominins and carnivores in the mix.

There were many unanswered questions and debates raged for decades, with researchers on both sides offering compelling evidence to support their arguments. Did hominins hunt small prey and scavenge larger ones? If they scavenged, did they force carnivores from their kills, or approach only after the bones had been picked clean and the predators left? More to the point, had the hominins purposefully used the sites as base camps for sharing, or did they unwittingly return to places because game concentrated there, or because they were natural safe havens with few carnivores to threaten them? There were so many ideas, and so many ways to interpret the evidence. In the end, though, the big question remained the same. Did hominins at Olduvai and Koobi Fora behave like living hunter-gatherers, or did they have some more primitive way of life, perhaps intermediate between their *Australopithecus* predecessors and modern human successors?

What's important for us here is that, whether the hominins hunted or scavenged, their diets changed in a fundamental way. We don't know whether meat and marrow were rare treats or regular staples, especially at first. Stone tools and cut-marked bones are found sporadically at sites dated to the onset of the Pleistocene, around 2.6 mya, and perhaps even earlier.[13] But the sudden appearance of large concentrations of artifacts and animal remains around two million years ago surely signals a change in the role of hominins in their world. Our ancestors had earned a place at the dinner table with the large carnivores, and meat and marrow were bountiful on their biospheric buffet, regardless of how often hominins chose to fill their plates with these foods.

This should sound familiar. We learned in chapter 2 that dietary adaptations in primates are more about kinds than proportions of items eaten. Recall the lemurs, mangabeys, and gorillas, each with pairs of species having adaptations for the same types of food even when eaten in different amounts. Tooth shape alone does not tell us whether mangabeys eat hard nuts or gorillas take wild celery every day or only during crunch times when other foods are unavailable. Nature doesn't seem to care when selecting for dental adaptations. What matters is that teeth give a primate the dietary options it needs. Wouldn't the same hold true for the stone tools that gave hominins better access to meat and marrow?

But perhaps we need to look beyond lemurs, monkeys, and apes to interpret the evidence at Olduvai and Koobi Fora. After all, we're now talking about stone-tool-making hominins that butchered and ate other animals, sometimes animals larger than themselves. There's only one species of primate that does this today—us. It's time to shift gears and look inward, at least to those of us who still forage wild foods for a living.

Human Foragers

There are only a handful of cultures today that subsist by hunting and gathering, and few, if any, remain completely untouched by the outside world. These societies have been on the decline for a long time, and anthropologists have worked fast and furious to document their disappearing ways of life across the globe—in the Australian Outback, sub-Saharan Africa, tropical South America, and the higher latitudes of North America. Those studies were at their peak in the 1960s, when cultural anthropologist Richard Lee and evolutionary biologist Irven DeVore organized a discipline-defining symposium called *Man the Hunter* in Chicago. Their idea was to bring those who had done field-work among the remaining hunter-gatherers together with archaeologists and paleontologists interested in human evolution. Perhaps those few enduring foragers could teach us something about our collective past. Glynn Isaac was there, and the symposium surely helped inspire and shape his home-base hypothesis.

Man the Hunter

The take-home message that resonated loudest was that a sex-based division of labor with men hunting and women gathering was key to our ancestors' success. And while participants acknowledged that plant foods and gathering were important, the focus was squarely on meat and hunting. As the argument went, the mother-child bond is fundamental in mammals, and moms have been feeding babies for more than 200 million years. There was nothing new about that. But hunting and meat sharing brought dad into the picture. This is what paved the way to the human nuclear family. Women bore and cared for the children, and men provided for their nutritional needs. That would allow for a longer period of child dependency, and more time to learn the tricks of the trade from both parents. Hunting large game also meant cooperation between males, and the bonds and coalitions that developed led to a higher social order.

People's ideas of life in the early Pleistocene at that point looked a bit like the 1950s sitcom, *Father Knows Best*. By the mid-1970s, though, some began to wonder whether those ideas were more a reflection of the scholars who developed them (mostly men raised in the early twentieth century) than of the evidence itself. *Man the hunter* implicitly gave men the principal role in making us human. That just didn't seem right. Surely *woman the gatherer* was just as important to the story, and the complementary roles that men and women played in subsistence were the very root of the longevity and success of our family tree. But where was the evidence? There were plenty of butchered bones at Olduvai and Koobi Fora, but the archaeological record remained frustratingly silent on plant foods and the role of gathering in human evolution. There had to be another way of digging out the details.

Grandmothers and Tubers

Kristen Hawkes and her colleagues and students at the University of Utah turned once again to hunter-gatherers for answers. Hawkes started teaching at Utah in 1973. Back then she was focused on writing up her doctoral dissertation, a study of kinship in a tribe from New Guinea that grew sweet potatoes and raised pigs. This was classic cultural anthropology. But her work would soon take her in a very different direction. Hawkes was introduced to Eric Charnov, an evolutionary ecologist who, like her, had just moved to Utah from the University of Washington. Charnov understood foraging as a compromise, a balance between the cost of obtaining a food and the energy it yields. I think of Halloween, and my young daughters considering both "best" candy and distance between houses as they planned their trick-or-treat routes. Charnov developed much more sophisticated models with many more variables, but the general idea was the same. His models could be compared with real decisions made by real animals, including humans. Do they maximize return on investment? If so, how? If not, why not? The whole approach resonated with Hawkes, and she quickly realized its potential for anthropology.

Much of her work in the early 1980s focused on the northern Aché people of eastern Paraguay. One of Hawke's students, Kim Hill, had worked with the Aché when he was in the Peace Corps. The group he worked with was at the time transitioning from a nomadic hunting and gathering lifestyle to permanent settlement at a Catholic mission colony on the Jejui Guazu River just south of the Amazon basin. The Aché grew a few crops and raised some animals there, but families still trekked into the forest for days or weeks at a time to forage for wild foods. This gave Hawkes, Hill, and others the chance to follow the women and men as they gathered palm hearts and starch, wild fruits, and honey, and hunted armadillos, pacas, monkeys, and peccaries. Hawkes and her students documented the foods collected; counted their calories; and recorded time spent travelling, gathering, and processing edibles. When they plugged these data into Charnov's models, the results were much as predicted. Decisions the Aché made were largely a matter of costs and benefits.

There were some surprises though, at least for Hawkes. Another of her students, Hillard Kaplan, had been collecting data not just on the foods foraged, but also on who ate what. Men seemed to be targeting prey to share with the community at large rather than focusing efforts on feeding their own nuclear families. They gave an enormous amount of game and honey away. It was as if they were showing off to others instead of taking care of their wives and children, which didn't fit well with the prevailing man-the-hunter model. If the goal was to feed the family, there were better, more predictable ways to bring in calories.[14]

Just as Hawkes was beginning to come to terms with this realization, she and archaeologist Jim O'Connell were invited to study diets of another foraging society, the eastern Hadza. Unlike the Aché, many Hadza were full-time hunter-gatherers, and they often took larger prey. This would be a great opportunity to delve deeper into men's motivations and strategies for sharing meat. The Hadza turned out to be different from the Aché in many ways, which is not surprising given the differences in where they lived. For the Aché, it was rolling hills covered by broadleaf evergreen forest. Their lush and bountiful hunting grounds were crosscut with rivers and streams fed by more than five feet of rain a year. The Hadza, in contrast, lived in savanna woodland, with rocky hills and thorny acacias, myrrh, and baobab trees. And while downpours were common during the rainy season, Hadza got much less total precipitation throughout the year. There were seasonal

6.3. Hadza hunters heading into the bush near Lake Eyasi.

rivers and natural springs, but Hadzaland got pretty hot and dry before the rains came.

One curious and unexpected difference from the Aché was that Hadza children gathered much of their own food from a very young age, except in the late dry season when there was little fruit to be found. Is this where dad came to the rescue and brought home the bacon for his family? Hawkes and O'Connell set out for Hadzaland to see. They lived side by side with several camps over the course of 10 months and followed along as the men hunted and women gathered. They made detailed records on time spent foraging, energy yield, and food sharing, much as Hawkes and her students had done for the Aché.

Again, the men focused on large animals. But big-game hunting in arid East Africa is notoriously unreliable. The average rate of success for a given hunter on a given day is as low as three %. To be fair, the group success rate is much higher, and hunting parties make, and share, a large-game kill about once a week. But if a man's first priority is to his own wife and children, wouldn't it be more sensible for him to limit the risk and focus on smaller, more dependable game, or to collect honey and plant foods for that matter?

So what did the Hadza eat when there was no meat, and no fruit, to be had? The answer is buried just below the surface. There's an incredible reserve of energy beneath the savanna woodlands surrounding Lake Eyasi. Plants store carbohydrates and water underground in swollen tuberous roots and stems to protect their reserves from hungry herbivores. Those tubers are different than our potatoes and yams—much more fibrous. The edible parts offer only about half the calories of cultivated tubers, but they are available year-round, wet season or dry. More important, they deliver the energy Hadza need to get through lean periods and to supplement their diets at other times.

Young children don't have the skills, strength, or stamina to dig them out, though. That's mom's job or, when mom's busy providing for a younger brother or sister, it's grandma's. Hawkes and O'Connell wondered whether this might be a way for older women to ensure the successful spread of their genes after they themselves could no longer bear children. If grandmothers also provide for the young, their daughters can have more dependent children at one time. This might not be a big deal in the rainy season when they can gather fruit for themselves, but by late in the dry season when tubers are all there is, having a grandmother to help can mean providing for both a newborn and a toddler rather than just one child. Perhaps this is why human foragers can have children at twice the rate of chimpanzees, or three times that of orangutans. While the man-the-hunter model explained our unusual birth spacing by fathers bringing home meat to feed the kids, there were other possible explanations.

So Hawkes and O'Connell began to put together a new model for the evolution of human diet. Drier conditions in the early Pleistocene would have meant less fruit and more lean times. But rather than an increasing role for men and meat, perhaps it was grandmothers and tubers that were important. The idea that women and foraging were key to a more human way of life wasn't a new one, but Hawkes and O'Connell introduced a novel twist. They based their model on data from real people living little more than a stone's throw from Olduvai Gorge. These were the last of the hunter-gatherers to walk the very same plains as our hominin ancestors. They also stressed that the Hadza roast their tubers over an open fire. Hawkes and O'Connell have suggested that this helps break down toxins and increases digestibility. It also fits the "grandmothering" theme, as young children cannot make and tend fires by themselves. So cooking, too, must have been an important milestone in the evolution of human diet.

Cooking Made Us Human

Hawkes and O'Connell weren't the only ones thinking about cooking and human evolution, though. There are two copies of *Catching Fire* in my house. There's my daughter's, the second book of the *Hunger Games* trilogy. That one has little to do with food, or lack thereof for that matter. Then there's mine, the popular science book subtitled *How Cooking Made Us Human*, by primatologist Richard Wrangham.[15] It is Wrangham who has most assiduously championed the idea that the invention of cooking was central to human evolution.

Like so many of the other scientists we've considered thus far, Wrangham started out on a very different track. He was an aspiring zoologist back when he was in his teens, and took a gap year before college to help with a study of antelope behavior in Zambia's Kafue National Park. It was there that he, like so many of us, fell in love with Africa. And it was then that he became keen on understanding how habitat shapes social behavior. He began working with chimpanzees in 1970. He was Jane Goodall's assistant at the legendary Gombe Stream National Park, charged with documenting social relationships between siblings. But Wrangham soon noticed that the cycle of life in the forest affected social interactions in a big way; chimpanzee groups became small and scattered when ripe fruits were scarce. This inspired him to continue at Gombe for his dissertation research and to study seasonal changes in food availability and their effect on group size and structure.

Wrangham tried just about everything the chimpanzees ate during those early years. You might imagine bananas and grapes or apples and peaches when you think of ripe, fleshy fruits. But these have been bred over the course of millennia specifically for us to eat. Most wild fruits are very different. They are tough and dry, fibrous and bitter—hardly fit for human consumption. It didn't take Wrangham long to realize that it would be difficult, if not impossible, for us to survive on a chimpanzee diet. This was made all the more obvious to him a few years later when he and his wife, Elizabeth Ross, studied Mbuti pygmies in the Ituri Rainforest of what is now the Democratic Republic of the Congo. It was a chance to study the diets of people living in the same patch of forest as chimpanzees. Long story short, the people there hardly ever ate chimpanzee foods, even when they were hungry.

But the importance of cooking didn't occur to Wrangham until many years after that. He was sitting by the fireplace at his home in Massachusetts, organizing his notes for a lecture on human evolution. Then it hit him. As O'Connell and Hawkes noted, cooking softens food and releases its nutrients. It improves taste, detoxifies, and makes a meal easier to chew and to digest. And all living peoples today cook. Maybe we have to because our ancestors did it for so long that we've lost the ability to burn our fuel without giving it a head start before it hits our gut. Wrangham has spent much of the past two decades in the laboratory, detailing the effects of cooking on food. His conclusion is that we just can't survive on a chimpanzee diet because we have become adapted to eating cooked food. Yes, cooking is a cultural thing, but that doesn't mean it didn't feed back into our biology and lessen selective pressures that would otherwise have kept teeth big and guts complex. The same argument has been made for stone tools and meat. Why not cooking?

But how could it have contributed to making us human? Man the hunter and the grandmother hypothesis aren't just about meat and tubers, they're about forging social contracts between mates or between mothers and their daughters for sharing food and taking care of the kids. These were the sorts of things that might have led our ancestors down the path. Wrangham suggests that cooking could have meant a social contract between males and females too, not just for sharing food but for cooperation to protect the pile gathered by the fire pit from would-be thieves.

THE HARD EVIDENCE

It doesn't take more than a brief stay with the Hadza to understand that how we get our foods, divvy them up, and prepare them are all part of what makes the human species unique and special. We don't have to visit a farm or a city to know that our place in the larger community of life is fundamentally different from those of other primates. Human foragers hunt more and larger animals. Many dig deeply for tubers, and all use a bevy of tools to get and prepare food. They cook their food to make it palatable, chewable, and digestible. They gather their meals to share with their young and others to build and grow social bonds of a complexity unparalleled in the animal kingdom. These are some of the things that make us different, that make us human.

But how can we know when all this started, and with whom? Piles of cut-marked bones and stone tools dug from sediments in the Great Rift Valley and from caves in South Africa offer some hints of humanlike, or perhaps near-humanlike, behaviors two million years ago. But we can also look to the hominin fossils themselves for evidence of the fundamental changes that made us human. And what better place to start than brain size? We have grossly enlarged brains, at least compared with other primates,[16] and it's not a stretch to think, with those grossly enlarged brains, that this was an important part of the package leading our ancestors into the human adaptive zone. Indeed, the renowned Scottish anatomist Sir Arthur Keith argued back in the late 1940s for a brain-size threshold, a "cerebral Rubicon" of sorts that marked the point of no return on the path to humanity. He suggested a cranial capacity of 750 cm^3, which falls halfway between the largest known values at the time for a gorilla and the smallest for a normal human.

Truth be told, we're not nearly as sure today of how, or even whether, cranial capacity translates to humanness. The recently discovered "hobbit" hominin, *Homo floresiensis*, was found alongside stone tools, animal bones with telltale marks of butchery, and perhaps even evidence of controlled fire.[17] Its cranial capacity is only half that of Keith's Rubicon—well within the range of living apes. Remember also that Louis Leakey and his colleagues trashed Keith's Rubicon back in 1964 by including *H. habilis* in the genus. Finally, variation within early hominin species can be extreme, with overlap between species and even genera, which muddles the idea of an inexorable increase (gradual or abrupt) in brain size from one species to the next. Nevertheless, there is a definite trend over time, culminating so far in our brain, which is really big by comparison to that of any other primate. This has important implications for our adaptive zone, if for no other reason than that a big brain is very expensive to grow and maintain.

Expensive Tissues

Our brains weigh nearly five times what you'd expect for a mammal of our size. That's the difference between an apple and a pineapple, and it means a lot in terms of our fuel requirements. A human body at rest uses about as much energy as a typical household lightbulb, something like 60 watts. Our brains account for a dozen of those watts, about 20% of our daily expenditure. That may not sound like much, but when you consider that the human brain makes up only 2% of our body weight, it is really very impressive. In other words, our brains burn calories at a rate nearly 10 times that of the body as a whole. It takes a lot of energy to fire nerve cells and pump ions across cell membranes. Because energy used depends on tissue mass, our relatively large brains require an extraordinary amount of fuel.

It may come as a real surprise, then, to learn that humans don't burn any more calories in a given day than expected for a mammal of our weight. It's a real conundrum. How can we balance our energy budget given our large expensive brain? This question was very much on Leslie Aiello's mind in the early 1990s. Her work at University College London at the time focused on relationships between the overall size of the body and the parts that make it up. The energy requirements of the individual bits and pieces are important to understanding how their sizes relate to the energy requirements of the body as a whole.

Aiello reasoned that because the energy required to run an organ depends on its mass, the cost of a larger brain must have been offset by a smaller something else. At first, Aiello thought it was our wimpy muscles. Chimpanzees are four times stronger than us. Try arm wrestling one sometime; they have incredibly powerful upper limbs to propel them through the trees. Could our upright-walking ancestors have sacrificed muscle mass to shunt power to their bigger brain? No, that wasn't it. Even though our muscles account for nearly half our body weight, they only represent 15% of our energy expenditure. There had to be something else that accounts for the energy balance.

Aiello enlisted the help of a physiologist from Liverpool John Moores University, Peter Wheeler, and the two turned their attention to more "expensive" tissues, like the heart and abdominal organs. Wheeler at first thought it might be the kidneys. After all, our ancestors evolved in arid conditions. Maybe we didn't need large kidneys to produce a lot of urine. But, again, that wasn't it. Our kidneys are no smaller than expected. The same goes for our liver and our heart.

Our gut, on the other hand, is smaller. In fact, when you add our gut and brain together, their summed weight is just about what we'd predict for a mammal our size. It looks as though the energy saved by a smaller gut compensates for that required by a larger brain. This is actually not uncommon in primates. Folivores tend to have smaller brains and larger guts than closely related frugivores. Think about it. Leaves require larger guts for digestion, but a huge brain isn't particularly important when the world is your salad bar. Maybe you prefer younger, more succulent ones, but leaves are everywhere in the canopy. A frugivore, on the other hand, doesn't need an elaborate gut to digest fruit flesh, but it takes more brain power to understand the vagaries of food availability in time and space.

So Aiello and Wheeler proposed that as hominins evolved larger brains, their guts got smaller to compensate and maintain the energy balance. That meant a change in diet to foods that might require more brain to acquire but less gut to digest. As our ancestors evolved from *Australopithecus* to the earliest *Homo* and on to *H. erectus*, foods that challenged the brain, but not the gut, must have become increasingly important. To Aiello and Wheeler, the most logical culprit was meat, though less fibrous tubers, especially cooked ones, might have done it too. That said, the "expensive-tissue hypothesis" has recently been called into question, for mammals as a whole anyway,[18] and we've taken the argument about as far as we can in this discussion. It's time to move on to the teeth. Maybe they can help us sort through the various ideas about diet changes that led, or followed, our ancestors into our generic adaptive zone.

Fossil Homo *Teeth*

There are some challenges to understanding form and function of early *Homo* teeth that we didn't have to worry as much about with *Australopithecus* or *Paranthropus*. First and foremost, the dental evidence for this critical point in human evolution, some would say *the* critical point, is surprisingly meager. All of the *Homo* teeth from the early Pleistocene of Africa— *H. habilis, H. rudolfensis*, and *H. erectus*—would fit in a shoebox, and a small one at that. In fact, there aren't much more than a couple of dozen specimens each identified as *H. erectus* and either *H. habilis* or *H. rudolfensis* (the two can be difficult to separate).[19] And many of these are isolated, individual teeth. So for any given tooth type, say the lower second molar, there are at most a handful for each species, some of which are broken, worn, or look as though they've been through a rock tumbler. It's difficult to be confident of our interpretations when there are so few good fossils to work with.

Also, if tools began to take on an increasingly important role in food processing, the rules of the game relating teeth to diet might have started to change with early *Homo*. Recall from chapter 1 that teeth are about food fracture properties, not food type per se. If hominins used tools to soften hard foods or tenderize tough ones, relationships between dental form and food type might have started to blur.

Let's consider how living animals use tools today. In most cases, they provide access to foods that would otherwise be off the table. Egyptian vultures crack hard eggshells with rocks. Crows, orangutans, and chimpanzees fashion wooden probes to forage for insects. Dolphins use conch shells to trap fish. Sea otters and macaques use hammerstones to break clam and oyster shells. Capuchins use rocks to smash open nuts. The list goes on and on. There are occasional exceptions, like the chimpanzee at Mahale Mountains National Park in Tanzania reported to stick twigs up its nose to trigger sneezing. Even in this case, though, the animal eats the resulting mucus.[20]

For our purposes, though, tools are largely about increasing options on the biospheric buffet. But in doing so, tools can change the material properties of foods that teeth have to contend with. Also, the chances are good that the flaked stones from Olduvai and Koobi Fora were just the tip of the iceberg. We can't see tools made of wood or other plant parts in the very early archaeological record, but we have to assume that they were just as important to hominins as they are to apes today, if not more so. In other words, stone flakes were probably just part of an expanding kit of tools used more and more for obtaining and preparing foods.

Tooth Size

Back when Louis Leakey and his colleagues announced the discovery of Jonny's child, they argued that as tools became more important, teeth became less so. Processing foods outside the mouth would have taken pressure off the teeth, allowing them to become smaller. So big brains and nimble hands would have gone together with little molars and wimpy jaws as one neat package in early *Homo*, as the toolmakers were remade by their tools.

But when we sit down with the teeth, measure them, and compare their sizes to all the other hominins, the story doesn't quite add up. Sure, *H. habilis* molars are small compared with those of *Paranthropus* found alongside them, but when we compare them (and those of *H. rudolfensis*) to *Australopithecus*, they aren't small at all, especially if we take body size into account. In other words, the earliest members of our genus hadn't actually evolved smaller back teeth, *Paranthropus* had evolved larger ones. In fact, cheek teeth didn't get much smaller until *H. erectus*, whose first known appearance was more than half a million years after the earliest stone tools. So if tools led hominins to smaller molars, it took a while.

The front of the mouth is a different story, though. As best we can tell, *H. habilis* and *H. rudolfensis* both had larger incisors for their body sizes than *Australopithecus* or *Paranthropus*. As we learned in chapter 3, the difference between *H. habilis* and *Australopithecus* in incisor size relative to molars is about the same as that between orangutans and gorillas. Since orangs husk fruits with their front teeth, whereas gorillas are adapted to grind more leaves and other tough plant parts with the back ones, it might also be that early *Homo* used its incisors more than *Australopithecus*, not less. *Homo erectus*, though, did have smaller incisors than those of *H. habilis* or *H. rudolfensis*, about the same relative size as *Australopithecus*. It may be, then, that *H. habilis* and *H. rudolfensis* ate foods requiring more preparation outside the mouth, but that it wasn't until *H. erectus* that tools really began to relieve the front teeth of their duties.

It's hard to say, though, what these changes in tooth size actually meant. Not only are our sample sizes small, but our body weight estimates are also terribly imprecise. We can't get a good sense of what tooth size actually means without an idea of the size of the body the teeth are attached to. It's all relative. Think about your teeth in a mouse's mouth, or an elephant's. And even if we had museum vaults filled with complete skeletons, I'm not sure how much we could coax about diet from tooth size. Yes, species with larger front teeth tend to eat foods that require more incisor use, like husked fruits. But the relationship between molar size and diet isn't that clear. Monkeys in South America that chew a lot of tough leaves tend to have larger molars than fruit eaters, which makes sense, but we see the opposite pattern for monkeys in Africa and Asia. That's because there's more to molar size than the amount of chewing— space in the jaw, for example. It's all rather difficult to sort out.

Tooth Shape

Molar shape seems to work better. In chapter 1 we learned that gorillas and other primates adapted to eating tough foods, like leaves and celery stalks, tend to have steeper, rougher biting surfaces than fleshy-fruit eaters, like chimpanzees. Mangabeys and others adapted to hard objects have especially blunt, flat cheek teeth. These rules of thumb hold for all groups of primates, whether they're from South America, Africa, or Asia.

So what do we know about tooth shape in early *Homo*? Phillip Tobias described *H. habilis* back teeth as having high, sharp cusps covered in thin enamel, at least when compared with *Australopithecus*. *Homo habilis* didn't wear them flat like *Australopithecus* did either. But what happens when we put numbers on tooth shape? Can we separate early *Homo* species from their *Australopithecus* predecessors? Recall from chapter 1 that we can measure tooth cusps and fissures using a laser scanner and GIS tools designed for modeling mountains and valleys, and that we can compare our measurements for modern species to those of early hominins at similar stages of wear. When we do this, early *Homo* does separate from *Australopithecus*. Samples differ about the same amount as chimpanzees and gorillas, with early *Homo* between the two ape species and *Australopithecus* with flatter surfaces than either.

No paleoanthropologist with a sharp eye and a keen mind would argue that early *Homo* had especially sharp molars. Still, their back teeth would have been better suited for shearing tough foods than would those of their hominin predecessors, *Australopithecus*. In contrast, *Australopithecus* would have been able to crush hard foods with less risk of breaking their teeth. This suggests a change in diet with the earliest members of our genus, perhaps with increasing emphasis on meat or fibrous plant parts. But there are a couple of issues to consider before we take this interpretation too far. First, again, our sample sizes of early *Homo* are tiny because we can only compare similarly worn teeth at the same position in the mouth. In fact, I had to combine all available lower second molars of *H. habilis*, *H. erectus*, and *H. rudolfensis* into a single sample just to have enough to compare with *Australopithecus*.[21] That means no comparison was possible between early *Homo* species themselves.

The other issue is that, even if preparation with tools hadn't changed food properties beyond a tooth's recognition, we need to remember that tooth shape is about kind, not proportion (see chapters 2 and 5). In other words, slope or jaggedness of a crown can tell us something about what a hominin was capable of eating, but not what it ate on a daily basis. Early *Homo* could have eaten the same sorts of foods as its *Australopithecus* predecessors most of the time—or not. That is, as we learned in chapter 5, where foodprints—the chemistry and microscopic wear of teeth—come in.

Foodprints

The mix of light and heavy carbon in the teeth of early *Homo* is about what you'd expect for a diet that includes both tree or bush parts and tropical grasses or sedges. We can't tell whether that carbon came directly from plants or from animals that ate those plants, though. What we can say is that the early *Homo* biospheric buffet was stocked with foods derived from both tropical grasslands and forests. In this way, early *Homo* carbon isotope ratios aren't much different from those of *Australopithecus africanus* or *Au. afarensis*. There aren't differences between South African and eastern African early *Homo* carbon isotope ratios either, as there are for *Paranthropus*.

The microwear tells a slightly different story, both because *H. habilis* is different from *Australopithecus*, and because *H. erectus* is different from *H. habilis*. The average *H. habilis* microwear surface is similar to that of *Australopithecus*, though the *Homo* species shows a greater range of variation, from surfaces covered in light scratches to those with a bit more pitting. *Homo erectus* teeth vary even more, with some microwear surfaces up into the range we'd call substantially pitted. This all suggests that while none of these species were likely hard-object specialists, *H. habilis* may have eaten a bit more hard food than did *Australopithecus*, and *H. erectus* may have eaten even more. If the variation we see in microwear texture can teach us something about diet breadth of a species, *H. habilis* may well have had a more flexible diet than *Australopithecus*, and *H. erectus* may have had a more versatile one than *H. habilis*.

ASSEMBLING THE PIECES

So that's our evidence: tooth size, shape, and wear; models based on living foragers; and the early archaeological record. The story was reasonably straightforward back when Jonny's child was first found. Stone tools meant a new way of life and access to foods that would otherwise have been off the table. It was our ancestors' way of coping as savanna overtook their ancestral home and forest resources began to dry up. Teeth shrank, hands became more dexterous, and brains expanded as the toolmakers were transformed by their tools. The end result was the human genus.

Homo *Whodunit*

But new evidence and interpretations have suggested that the story wasn't quite so simple. *Homo habilis* and *H. rudolfensis* molars weren't really much different in size than those of *Australopithecus*. Bernard Wood of George Washington University and Mark Collard of Simon Fraser University have even argued that *habilis* and *rudolfensis* don't belong in *Homo* at all, and should be relegated to *Australopithecus*, whether or not they had tools.[22] On the other hand, they did have bigger front teeth, so there may well have been changes in diet, with selection for foods that had to be separated from inedible parts before they were chewed, like fruits protected by husks or tendons attached to bone. Also, early *Homo* molars were somewhat sharper and "crestier," meaning more efficient slicing or shearing of tough foods. And the microwear suggests that *habilis* was less picky than *Australopithecus*.

The differences between *H. erectus* and *Australopithecus* were more extreme and obvious. First, *H. erectus* does have substantially smaller molars than *Australopithecus*, and its front teeth are also smaller than those of *H. habilis* or *H. rudolfensis*. Maybe it was really with *H. erectus* that tools began to relieve the selective pressures on teeth and jaws that kept them big. Indeed, we start to find large concentrations of stone artifacts and cut-marked bones around two mya, just before the earliest record for *H. erectus*. Perhaps this in part helps explain how *H. erectus* was able to maintain its larger brain. If so, our ancestors' climb to the human adaptive plateau was at least a two-step process involving first *H. habilis* or perhaps *H. rudolfensis*, and then *H. erectus*.

Of course, we don't know for sure who made the tools found at Olduvai, Koobi Fora, and the other early archaeological sites. Without that, our inferences about the hominins that used them are largely speculation. Remember that we've got at least four species wandering about the landscape between 2 and 1.5 mya. We are probably on solid ground with *H. erectus* making and using stone tools, though. As best we can tell, only they made it out of Africa in the early Pleistocene, and sites in Eurasia have concentrations of stone tools and cut-marked bones. But what about *H. habilis* and *H. rudolfensis* or, for that matter, *Paranthropus* or even *Australopithecus*? The earliest tools preceded *H. erectus* by more than half a million years. It's reasonable to speculate that *H. habilis*, at least, also made tools. That would have given its opposable thumbs something to do. We can't exclude the others either.

Changing the Game

We've learned from our studies of nonhuman primates that food choice is a matter of availability, and that this depends both on what there is to find in a given place and at a given time and on what teeth and guts allow a species to gather, process, and assimilate.

Early *Homo* had little control over what nature had to offer, and options on the biospheric buffet must have changed all the time. The Pleistocene was an epoch of climatic instability, with shifting landscapes and habitats. We learned in chapter 4 that this wasn't merely grassland overtaking forest. Remember that carbon isotope ratios of *H. habilis* and *H. erectus* indicate a mix of open- and closed-country foods, much like we find for their late-Pliocene *Australopithecus* predecessors. That didn't seem to change. The bigger story was the increasingly variable climate with alternating cycles of warm-wet and cool-dry conditions. Hominin habitats must have fluctuated as lakes expanded and contracted with the spread of the Great Rift Valley. Forest gave way to grassland, then grassland to forest, as the Earth wobbled on its axis and its orbit went from circular to elliptical and back. In this light, increasing microwear variation from *Australopithecus* to *H. habilis* to *H. erectus* makes sense. So does an increasing role for tools to gather and process a broader spectrum of foods. That could mean more options in a given place and time. Maybe this gave early *Homo* the versatility needed to weather the storm in its increasingly unpredictable world. Maybe that is the key to the human adaptive zone.

We are fundamentally different from the other animals on our planet. To understand how, we need to strip away the cities and villages, the farms and ranches. Our ancestors were human well before any of those. Savanna stretches across Hadzaland as far as the eye can see from the top of Gideru Ridge. From that vantage point it becomes clear that we are human by virtue of our unique relationship with the larger community of life. Tools are an important part of the story, as are cooking and sharing animal prey and gathered plants. These are the sorts of things that must have led our ancestors out of Africa and allowed them to find sustenance wherever they went. Their approach to gathering, processing, and distributing food let them make the most of what nature had to offer.

But, for some, that wasn't enough—they began to grow crops and raise animals. Our next stop is the Neolithic Revolution. We can view it, too, through the lenses of teeth, diet, and a changing world.

CHAPTER 7

The Neolithic Revolution

Planting seeds and raising animals changed our relationship with the world around us in a fundamental and profound way. They drove our ancestors from being a part of nature to being apart from it. As the renowned Australian archaeologist V. Gordon Childe wrote, "with the adoption of agriculture and stock raising, [man] became a creator emancipated from the whims of his environment."[1] The *Neolithic Revolution*, as Childe called it, was a milestone of biblical proportions. Think about it. Stable resources and surpluses are the fuel of civilization. Food production meant that our ancestors could gather in large communities and form complex societies. Once they started to cultivate and domesticate plants and animals, it didn't take long for the first great cities to spring up in places like Mesopotamia, India's Indus Valley, and along the Nile in Egypt.

Before the Neolithic Revolution, we humans played by the same rules as everyone else. For nearly 200 millennia, people were limited to the items nature chose to set out on the biospheric buffet. They foraged for wild foods, just as earlier hominins had for millions of years before them. But when the environment changed the food options did too, and our ancestors had to respond. It wasn't easy. Some moved, some died, and some adapted. Recall Elisabeth Vrba's Hindu triad analogy in chapter 3—Vishnu the preserver, Siva the destroyer, Brahma the creator. Such responses drive evolution. They explain how our changing world has given countless species their own unique identities, and how it made us human.

But our ancestors changed the rules of the game. They started raising animals and growing plants—rye, wheat, and barley in the Near East; rice and millet in China; sorghum in Africa; corn, potatoes, and squash in the Americas; taro and bananas in New Guinea. Humans had earned a living by hunting and gathering wild foods for 10,000 generations, but in just a few, brief millennia, food production sprung up across the globe. It happened separately in at least a dozen places. Dominion was ours, and it began a cascade of events that gave humanity its greatest accomplishments, from the peanut butter sandwich to the deep-space probe.

When, where, and why did this revolution, the one that changed everything, happen? Was it an inevitable outcome of our being human, with all the social and political baggage that comes with that? Did we outgrow nature's ability to provide for us on its own? Or were we driven to agriculture by global climate change at the end of the Pleistocene epoch, the last great ice age? Archaeologists have dug deep to uncover the details of our transition from foraging to food production. Yet nearly a century of work has brought more questions than answers. The only thing most agree on is that there are no simple answers.

In this chapter we set our sights on the Neolithic Revolution, looking forward from the distant past and considering models through the lens of climate change, its effects on food availability, and the choices our ancestors made to earn a living. We visit two of the most important sites in the Near East, Ohalo II and Abu Hureyra, and meet the archaeologists who dug them and gleaned insights from the remains they found. We then head to cold and barren central Greenland and open the time capsule that is its immense ice sheet, an archive of our planet's climate history. And we end our journey back in the lab, with the teeth and bones of those who witnessed climate changes and the transition from foraging to food production. How have these things affected human biology? We consider familiar foodprints, along with some new ones that offer important insights into our evolution, both past and present.

AN OASIS IN THE DESERT

Climate change clearly played an important role in human evolution. But what has it done for us lately? Childe believed that our ancestors were driven to cultivation and domestication of wild plants and animals by their changing world, too. He speculated that rain patterns must have shifted when the glaciers that covered much of Europe melted at the end of the last ice age. That would have left northern Africa and western Asia parched in the process. As grassland gave way to desert, the people and animals that lived there would have been forced together into places that were still well watered, like the Nile, Indus, Tigris, and Euphrates valleys. It was no coincidence to Childe that these were precisely where the wild ancestors of wheat, barley, sheep, and oxen were found, and where the earliest farmers and pastoralists had domesticated them.

It's not surprising that textbooks usually credit Childe with the idea that the end of the Pleistocene triggered the Neolithic Revolution. His was among the biggest names in twentieth-century archaeology. But it's a shame, too, because students rarely get to learn about Raphael Pumpelly, the geologist-explorer and adventurer who *really* came up with the oasis theory. His is a tale worth telling. Pumpelly laid out the argument during his presidential address to the Geological Society of America in 1906, the year before Childe entered grammar school.

It had taken him four decades and some incredible twists of fate to assemble the pieces. The story is a bit like Forrest Gump meets Indiana Jones, with Pumpelly witnessing history as he made his way across continents from one adventure to the next.[2] He began his career as a geologist establishing mines in Arizona just before the Civil War, but was forced to flee westward, escaping Apache warriors and Mexican bandits on horseback. He rode on to California, and then sailed for the island of Hokkaido to work for the Japanese feudal government to help develop its mining industry. But Pumpelly had to make a quick exit from there too. The emperor was using antiforeign sentiment to wrest power from the ruling shogunate, and civil unrest was brewing. He again headed west, returning home through mainland Asia, Europe, and across the Atlantic. His trip around the globe took nearly five years.

While Pumpelly was in China, an old map of the Tarim Basin in Xinjiang Province caught his attention. It was in a book of writings by Confucius and had a note on it that read, "Here dwell the Usun, a people with red hair and blue eyes." Another note mentioned vast rivers and ancient cities engulfed by desert sands in the distant past. Other maps referred to the Gobi desert as "Han-hai," dried sea.[3] Could there once have been a vast inland sea across central Asia? Perhaps the Aral, the Caspian, and the countless small lakes dotting the area today were its last remaining remnants. Pumpelly thought of Louis Agassiz and his theory, which was gaining popularity at the time, that much of the northern hemisphere, as far south as the Mediterranean and Caspian, had once been covered in glacial ice. Perhaps ice and snow had fed that inland sea, but the area dried up once the ice age ended. And what about those red-haired, blue-eyed people that lived along its shoreline? Could they have been the original Aryans that eventually spread westward to Europe as desert overtook central Asia?

It was a wonderful and novel idea. But Pumpelly had to put it aside when he got home. He had a new job at Harvard and a lot of geological survey work to do in the United States. He would turn back to central Asia, but not for another three decades, when he learned shells had been found in ice-age deposits there. It seemed there had indeed been a vast inland sea. This led him in 1903 to what is now Turkmenistan, where he found, in the foothills of the Kopet-Dag Mountains just south of Ashgabat, the archaeological site of Anau. His team dug the site the following year and uncovered bones of deer, gazelle, and other wild animals in the oldest layers, but oxen, sheep, and pigs farther up. They found the remains of wheat and barley at the site too.

Pumpelly composed his story. It seemed that the great inland sea, which had covered much of central Asia, began to recede at the end of the ice age. As deserts began to form, people congregated in oases, including Anau, which was fed by water flowing downslope from mountains to the north. As the population grew, increasing demand for food required that seeds be planted and animals raised. While most archaeologists at the time believed that Mesopotamia or Egypt was the cradle of civilization, for Pumpelly it was central Asia, and environmental change that marked the end of the Pleistocene had triggered the first step, the shift from hunting and gathering to herding and farming.

Pumpelly's oasis theory gives us a vivid picture of how science and history can intersect with serendipity and lead to innovative and compelling ideas. But, unfortunately, time has not been kind to Pumpelly or his theory. While central Asia was surely wetter than today, and the Aral and Caspian were larger,[4] there was no great inland sea during the late Pleistocene. Besides, people were planting crops and raising animals in the Near East thousands of years before they arrived at Anau. Pumpelly got most of the details wrong, which is probably why none of the archaeologists in my department had heard of him when I asked them. But his basic idea has survived: the shift from foraging to farming was a direct result of climate change to warmer, wetter conditions at the onset of the Holocene epoch around 11,600 years ago (BP).[5] This inspired V. Gordon Childe and, through him, many archaeologists today. But not all.

CHANGING OUR MIND

The ice sheets that covered much of the northern hemisphere have come and gone a hundred times since our earliest *Homo* ancestor made its first appearance. Through all but the last of the great glacial retreats, humans have clung to the same basic routine, hunting what animals they could and gathering the wild plants around them. To be sure, they got better at it over time by inventing more sophisticated tools for obtaining and processing food. These gave them more options to meet their nutritional needs and allowed them to spread across the planet. But until very recently, people were still content with what nature had to offer, despite the unpredictable, ever-changing menu. If the Neolithic Revolution resulted from climate change, why didn't it happen earlier? There were countless opportunities.

Some say it's because the Neolithic Revolution wasn't about climate change at all; it was about a change in the way people viewed their place in nature.[6] People started to think of themselves as apart from nature rather than part of it. Recall from chapter 6 the Hadza hunter-gatherers up on Gideru Ridge. When I asked them how we were different from other animals, they struggled for an answer. One elderly man wondered out loud whether it might be that humans have less hair. A young woman added that perhaps it was that we walk on two legs. I got no sense that the Hadza think there is a fundamental difference, let alone that they see themselves as having dominion over nature.

Perhaps, then, the Neolithic Revolution *was* about coming to see ourselves as masters of the natural world. Some archaeologists believe we can see this in the relics that early farmers left behind, the mothergoddess and the bull, representing female fertility and male virility, respectively. They see these as symbols of a mindset, a glue of sorts that held together the large, permanent communities that made farming necessary. Maybe cultivating plants and herding animals were not a consequence of climate change, but an outcome of a different way of thinking about ourselves, one that some have called a new "psycho-cultural" paradigm.

Others have argued that the Neolithic Revolution was more about human nature, the drive for prestige and wealth. I think of Kristen Hawkes's show-off hypothesis (see chapter 6). She believes Aché hunters purposely target large game to share with the community, rather than going after smaller, more reliable prey to feed their own families. Perhaps some just had too much ambition for the lifestyle of an egalitarian hunter-gatherer. The go-to example is the traditional potlatch of the indigenous peoples of the Pacific Northwest coast. "Big men" worked hard there to accumulate surpluses so they could throw huge parties for neighbors and friends.[7] They gave away food and other valuables to gain social status. Maybe settling down to provide for one's needs and then some, rather than climate change per se, drove the Neolithic Revolution.

THE HARD EVIDENCE

To be honest, though, I'm not a fan of the psycho-cultural or big-man models. It's not that they aren't elegant and compelling explanations. The problem is that they're difficult to evaluate. We simply cannot get into the minds of people who lived 10,000 years ago. Yes, there is evidence for sedentism and storage of food, but what do these really teach us about the motivations of people who lived in the past? Even if we could document community-wide feasts, how could we tell whether they were potlatches or potlucks? Also, these models still don't explain why it took nearly 200 millennia to make the transition to agriculture. Then there's the chicken and egg problem. Did a change in the way our ancestors viewed their world come first, leading to cultivation and domestication, or did food production itself inspire the change in their mindset? Did ambition drive some to plant crops, or did cultivation come first, providing farmers with what they needed to make and climb the social ladder? The archaeological record gives us great snippets of the past, but not enough to answer these very fundamental questions.

Speaking of chickens and eggs, back when I was in graduate school, the big debate was whether the needs of growing populations pushed people to plant crops and raise animals, or whether the surpluses that came with agriculture pulled communities to grow. Some argued that population growth is natural, but kept in check by the land's carrying capacity.[8] The number of people the biospheric buffet can feed is limited by nature's ability to keep it stocked. Perhaps the drive to thrive, to be fruitful and multiply, might have been enough to force people to start producing their own food. That would lead them to settle down because crops don't move. It becomes a vicious circle. Settling down allows or even encourages people to make more babies, and growing populations need more food. Some say populations got too big to move so they had to settle, but there weren't enough resources at the core to handle them. Others think populations settled in lush areas, but had to expand into more marginal settings as they grew. Perhaps, then, a lack of natural resources around the fringes forced cultivation.

Countless tons of dirt have been moved by archaeologists over the past century in an effort to understand the transition from foraging to farming. Thousands of scholarly papers and books have been written to peddle one idea or another about what pushed or pulled us into the Neolithic. There is still much disagreement on why people started planting seeds and raising animals. It probably doesn't help that these things seem to have happened in different places at different times for different reasons. The resulting information overload makes it very difficult for us to make sense of it all. But we can narrow our focus in the hopes of figuring out what drove the revolution in a particular place. Let's consider the part of the Near East called the *Fertile Crescent*. This is where the Neolithic Revolution seems to have happened first, and perhaps where we have the best chance of connecting subsistence to climate change around the end of the Pleistocene.

The Fertile Crescent looks to me more like a boomerang. Some use the ancient civilizations of Egypt, Phoenicia, Assyria, and Mesopotamia to define it, from Egypt's upper Nile through to the eastern Mediterranean, or Levant, swinging up and around to trace the courses of the Tigris and Euphrates rivers as they flow to the Persian Gulf. Others extend the western arm only as far as the Sinai. Either way, it's bounded in the north by flanks of the Taurus and Zagros Mountains and in the south by the Syrian Desert. There are countless archaeological sites in the region that can give us a glimpse at early settlement, cultivation, and, ultimately, domestication.[9] They are our window on life before, during, and after the Neolithic Revolution. We can use them to explore what changed and what didn't, to begin to address the big questions there. Two sites in particular come to mind for me, Ohalo II, on the southwest shore of the Sea of Galilee, and Abu Hureyra, on the south bank of the middle Euphrates River.

Ohalo II

The Sea of Galilee is actually a lake, the largest in Israel. It's surrounded by about 30 miles of shoreline but that changes with lake level, which itself fluctuates with the annual rains. The level drops dramatically during drought years, given heavy demand on the lake as a source of drinking water for the country. It was 10 feet below normal back in 1989, and that exposed archaeological remains that had been submerged for a very long time, perhaps thousands of years. An amateur archaeologist found some animal bones, flint objects, and a human jaw on the beach about five miles south of Tiberias.

Dani Nadel, a keeper of prehistory for the Israel Antiquities Authority at the time, went in to investigate. It was just a few hundred yards away from Ohalo I, another site that had been discovered when the waterline was low three years earlier. Nadel had helped describe the first Ohalo site when he was a master's student at Hebrew University in Jerusalem. But the new site, Ohalo II, was clearly bigger and more important. Nadel quickly organized a short field season for fear that it would be flooded when the lake rose again. His field team was able to excavate a good chunk of the quarter-acre site when the water level allowed, between 1989 and 1991, and again between 1999 and 2001. They found the remains of huts and hearths, a human burial, and stone and bone tools. They sieved every trowelful of sediment that they dug and recovered hundreds of thousands of plant parts and bone fragments. The preservation was incredible—the best ever for a site of its age.

Ohalo II gives us an unprecedented snapshot of life in the Levant before the Neolithic Revolution. It was 23,000 years ago; the ice sheets to the north were near their greatest extent. Imagine wandering bands of fur-clad ice-age hunters stalking large game as they roamed the cold, dry steppe grasslands that would be the Middle East. Now forget that image. Ohalo II was nothing like that during the last glacial maximum. Envision instead a small encampment on the edge of a large lake with half a dozen huts walled with brushwood of willow, oak, and tamarisk and floored with grass mats. The people of Ohalo II lived there year-round, and thrived with the bounty that nature offered them in the lake and the woodlands around it. Nadel's team found charred remains of wild almonds, pistachios, olives, grapes, and many other plant foods. They found the bones of countless fish, but also gazelle, deer, hare, and other wild mammals. And they found a grinding stone in the first hut, its surface embedded with starch grains from wild barley, wheat, and oats that would have been ground into flour to make dough. There were sickle blades with the telltale gloss of use in harvesting, and weeds, the descendants of which today flourish in cultivated fields. These people weren't just fishing, hunting, and gathering fruits and berries. They were collecting wild grains, perhaps even planting some.[10]

What was life like during the last glacial maximum? Ohalo II gives us a new perspective, at least for the Galilee. People settled there as resources allowed. They gathered and maybe even grew cereals. It makes you wonder just how much of a "revolution" the Neolithic actually was. But that wasn't to come for more than 10 millennia. Our next stop is Abu Hureyra, about 270 miles to the northeast, in Syria.

Abu Hureyra

Lake Assad was filled in 1974, after completion of the Tabqa Dam on the Euphrates River. The dam was built to generate hydroelectric power and to make a lake that would provide water for drinking and irrigation. But it also meant flooding the middle Euphrates valley with two and a half cubic miles of water and inundating remains of ancient settlements along the original banks of the river. It would submerge Syria's national heritage and, in fact, the very cradle of civilization. This would be a big deal. Antiquities officials appealed to archaeologists in the international community for help excavating sites, salvaging artifacts, and preserving their precious legacy before the dam was completed. The Syrian government even adopted a law allowing foreign scholars to keep half their finds as an incentive for museums to fund the work.

The response was extraordinary—10 teams answered the call. They came from the United States, the United Kingdom, France, Belgium, Germany, Switzerland, the Netherlands, Lebanon, and Japan.[11] Andrew

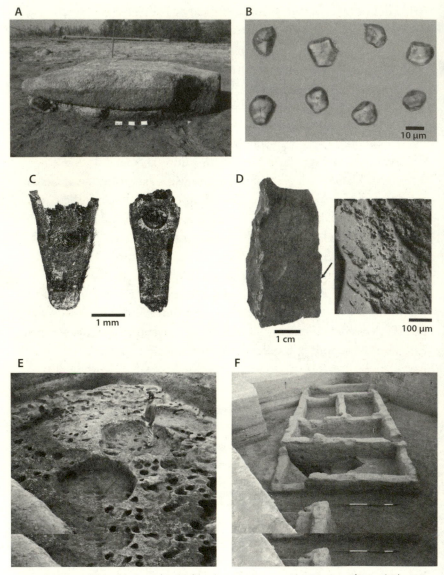

7.1. Remains from Ohalo II (A.–D.) and excavations at Abu Hureyra (E. and F.). A. a grinding-stone slab; B. oat starch grains; C. wild-type (*left*) and domestic-type (*right*) wild barley; D. a sickle blade (*left*) with use-wear polish (*right*) produced by harvesting (arrow indicates area magnified); E. Abu Hureyra I excavations; F. Abu Hureyra II excavations (A. courtesy of Dani Nadel; B. courtesy of Dolores Piperno; C. and D. from Ainit Snir et al., "The Origin of Cultivation and Proto-Weeds, Long Before Neolithic Farming," *PLoS One* 10 [July 2015]: e0131422, doi: 10.1371/journal.pone.0131422; E. and F. courtesy of Andrew Moore).

Moore, then a graduate student studying archaeology at Oxford, was among them. He had worked in Syria before, excavating with a French team at Tell Aswad near Damascus. The leader of that team put in a good word for Moore, and the Director General of Antiquities invited him to pick a site. It was 1971. Moore looked over the survey maps, visited the valley over the summer, and chose Abu Hureyra, a colossal occupation mound covering nearly 30 acres on the south bank of the Euphrates River. It turned out to be a good choice.

There was a lot to do, and Moore was short on time. It took the better part of a year to put together a plan of action and raise the funds he'd need to carry it out. That only left two field seasons before Abu Hureyra would be flooded, and it had more than a million cubic yards of archaeological deposits. Moore assembled a large team, up to 50 staff and workmen at a time, and got them started. But rather than bulldozing the site and digging as much as possible, he chose a more meticulous approach. He reckoned the devil would be in the detail. The team dug seven trenches in total, strategically placed to sample as many parts of the occupation mound as they could. They sieved every shovelful of dirt through meshed screens, and collected the tiny artifacts and bone fragments left behind. They also dumped thousands of pounds of earth into flotation machines filled with water. It was kind of like blowing into a glass of chocolate milk through a straw—a stream of air bubbles stirring up the mud to separate plant remains and charcoal from the sediment. These floated to the top and were scooped out and dried for analysis.

Moore's team dug for six months, two in 1972 and four in 1973. They found thousands of artifacts and hundreds of human burials, along with two tons of bone and more than enough plant debris, mostly tiny seeds and charcoal fragments, to fill a refrigerator. It was grueling work, and the tension built as the dam's completion date grew closer. To make matters worse, the trenches weren't yet deep enough to sample the lowest levels of occupation when the Arab-Israeli War hit, a few weeks into the second season. But time was running out. Moore pressed on—his team the only one in the valley that worked through the hostilities. There was little fuel to be had to run the generators and flotation equipment, let alone the stoves for cooking or vehicles to resupply the camp. They ran short of drinking water and food, and the army conscripted several of the young men. A couple of them came back in coffins. But the team persevered and, as a result, Abu Hureyra has produced an unprecedented picture of life during the transition from foraging to food production in the Fertile Crescent. No other site has given us such insight into the Neolithic Revolution, its causes, and its consequences.

Here's what we know. Abu Hureyra was occupied more or less continuously for 6000 years. Very few cities today can claim that kind of longevity. More important for us, its inhabitants witnessed the transition between Pleistocene and Holocene. At first they relied entirely on foods foraged from the wild, but later residents raised some of their own, and eventually became dependent on agriculture. What a great place to look for the causes and consequences of the Neolithic Revolution. Thanks to Moore's painstaking efforts, there's no better dataset. The plant and animal remains provide evidence of diet and habitat, artifacts and structures help us understand how residents met climate change with new technologies and lifestyles, and skeletons offer clues to the effects of those changes on the people themselves. We can piece together the story by matching the well-dated remains with paleoclimate data.

People first settled Abu Hureyra around 13,400 BP. We call these settlers *Natufians*.[12] The earliest village had no more than about 100–200 residents. It started with a cluster of small, round houses dug partly into the ground and covered with brushwood branches and reeds. People lived there year-round, and hunted and gathered in marshes and gallery forests along the river and in the woodlands that surrounded them. Their game was mostly gazelles, especially in the springtime when the herds came through, but they also hunted wild asses, oxen, sheep, and smaller animals. Plant foods were abundant too, at least at first. There were acorns and almonds, wild grapes and figs, peas and lentils, wheat and rye. There were wild melons, starchy roots and bulbs, and many other edibles growing within foraging distance from the village. This may not be the image most of us conjure up when we think of ice-age hunters, but why would anyone wander the open steppe in search of mammoths when a life of leisure in a land of plenty was just around the river bend? Natufians at Abu Hureyra settled down to enjoy the bounty, much as the people of Ohalo II had done 10,000 years before them.

Then wild plant foods began to drop out of the archaeological record, one by one. The biospheric buffet became increasingly sparse over the centuries. It started to thin out about half a millennium into the early occupation. At first there was a decline in more rain-dependent foods, like woodland fruits and nuts and wild lentils. Then wild rye and wheat grew scarce. Finally, even more drought-resistant plants, like club rushes and feather grasses, became less abundant. But just as wild foods waned, a few plump, domestic-like rye grains appeared, as did weeds like those found in cultivated fields today. Their numbers grew over time. The team also found grinding stone tools that inhabitants must have used to process the rye. There were lentils and domesticated wheat in slightly later levels too.

Then everything changed around 10,600 BP. After nearly 3000 years as a small village, Abu Hureyra was suddenly a Neolithic town of about 2000–3000 residents, and it would eventually grow to as many as 5000–6000. Envision hundreds of rectangular mud-brick homes and a lifestyle deeply rooted in agriculture. Countless weed seeds were sifted and floated from the upper levels of the mound; in places there were more domestic than wild cereal grains. Wheat and barley were the principal crops, though lentils were found throughout the sequence. Bread wheat became common later, as did chickpeas and field beans. These domestic cereals and legumes gradually replaced wild plant foods over time. There was also a shift from hunting gazelles and other animals to herding sheep and goats. Domestic cattle and pigs appeared toward the end too, albeit in lesser numbers.

ENVIRONMENTAL CHANGE AND THE NEOLITHIC REVOLUTION

Dani Nadel, Andrew Moore, and many others have spent decades gathering evidence for life in the Fertile Crescent before, during, and after the Neolithic Revolution. Ohalo II gives us a baseline, a picture of an encampment in the Galilee during the last glacial maximum. We don't see a hungry, cold band of nomadic hunters, ever wandering the barren steppe in search of their next meal. People lived there year-round, settled on the shore of a large, glistening lake fringed by rich woodlands. There were fish to catch, gazelles and deer to hunt, and all kinds of wild plant foods to gather. They may have even kept small gardens to cultivate wild-type barley, wheat, and oats. If life was supposed to be hard during the height of the last ice age, no one told the settlers at Ohalo II. To be fair, the occupation may have occurred during a warm and wet spike in the midst of the long cold and dry spell (see the following section). There must have been countless nomadic foragers living more challenging lives on the steppe to the north, too. Nevertheless, the evidence from Ohalo II foreshadows things to come and makes it clear that times were not tough for everyone throughout the last ice age.

Fast-forward 10 millennia to Abu Hureyra. While the abrupt change between the earlier and later occupations has led some to wonder whether there really was continuity between them, there is little doubt that the first inhabitants were hunter-gatherers and the last were farmers and herders. For the first half a millennium, residents were content to gather wild plants and hunt game in their lush river valley home. They settled well before they began planting seeds and raising animals. And if those rye seeds and weeds are really signs of cultivation, the population remained low for thousands of years after people there first started tilling the soil. It might be that farmers ramped up production because of population growth—consider the later settlement. But it doesn't look as though villagers were initially pushed to plant to support increasing numbers, at least not at Abu Hureyra.

So we're back to our original question. What motivated people to start growing food? Was someone trying to amass a surplus to throw parties and gain social status? Did they find gods and take dominion over nature? These are questions we can't answer. But maybe we can figure out whether climate played a role. To do this, though, we need a history of the Earth's climate that matches the resolution of the archaeological record. It's not enough to say that it was cold until the last ice age ended, then it got warm. We need more detail. We need to know how temperature and rainfall varied across the centuries, decades, and even years that spanned the Pleistocene–Holocene transition. Otherwise we have little hope of lining up diet or subsistence changes at sites like Abu Hureyra with climate. It's time to break open the archives of the Earth's recent climate history. Put on your parka—we're off to Greenland.

Ice (Sheet) Capades

Most of Greenland is covered in ice. The sheet extends almost 700,000 square miles across the island and is nearly two miles thick at its summit. It contains 10% of the world's total reserve of fresh water. If it melted, the sea level would rise 23 feet and inundate many of the world's coastal cities. This massive sheet of ice started forming about 150,000 years ago and has been building most of the time since. With each passing year snow piles on, and the layers below buckle under the weight of the ever-increasing load. Snow turns to what is called *firn* beneath the surface as ice grains are pressed together by the pressure. Deeper down, a couple of hundred feet below the surface, firn turns to ice as its air bubbles are sealed off. The process can take centuries. Air, water, and anything that settles on the fresh powder, like sea salt, dust, or volcanic ash, becomes trapped and buried. Greenland's ice cap is like a giant time capsule.

Paleoclimatologists have put a lot of effort into probing that capsule. In fact, much of what we know about the Earth's climate history over the past 100 millennia, especially for the northern hemisphere, comes from the Greenland ice sheet. Several teams have dug pits and drilled cores into the ice over the past half century to unlock its secrets, though two projects in particular, the Greenland Ice Core Project and the second Greenland Ice Sheet Project, stand out. These were separate studies conducted in tandem from 1989 to 1992 and 1993, respectively, the first by a team of Europeans the second by Americans. Their cores were sampled less than 20 miles apart, near the summit of the ice sheet in the center of the island. Both drilled nearly two miles from surface to bedrock, and they produced reassuringly consistent results.

Richard Alley of the American team recounted the experience in *The Two-Mile Time Machine*.[13] Ice-core drilling in Greenland is a huge production. In many ways it's similar to ocean drilling (see chapter 4), except that it's −20°F, the Sun never sets, and your laboratory is a trench cut into the snow. The American project involved two six-week-long legs per summer, each with a team of 50 drillers, scientists, and support staff. The working end of the rig was a spinning metal pipe about six inches across with teeth cut into its tip. It was suspended from a 100-foot tower extending from a geodesic dome that kept the elements at bay. The drill team worked day and night in shifts, cutting and bringing up the five-inch core three feet at a time. The science team put in 10 hours a day, six days a week, to study and prepare the sections for transport home. It took five years to reach bedrock and finish the job.

7.2. Midnight sun over the Greenland Ice Sheet Project 2 drill tower on the Greenland ice sheet. Photo courtesy of Richard Alley.

The ice cap summit is far north of the Arctic circle, so the Sun never sets in the summer or rises in the winter. During the summer months, surface snow is warmed by the Sun, so it's lighter and airier than winter snow. This causes layering that you can see in the cores, which can be used to count back through the years, like adding up tree rings. The layers get thinner and more difficult to distinguish with depth and the mounting pressure of the overlying ice. But the chemistry of the snow also varies between seasons, and blips in an electric current passed through the ice can be read to mark the years, too. Ages can be checked and corrected as necessary by matching them with the chemical signature of ash layers from specific volcanic events with known dates, like the eruption of Mount Vesuvius that buried Pompeii.

We can learn a lot about past climate from an ice core. Let's start with the water (H_2O) itself. We learned in chapter 4 that more than 99% of oxygen atoms have eight protons and eight neutrons (^{16}O), and that most of the rest have 10 neutrons (^{18}O). Recall that we can use the ratio of these isotopes in calcium carbonate nodules or foraminifera shells as a sort of "paleothermometer" to measure temperature in the past. We can consider water in ice the same way. In this case, the atomic weight of water molecules varies because of both oxygen and hydrogen isotopes. Deuterium (^2H) has a neutron, whereas the much more common protium (^{1}H) does not. Lighter water ice tends to form from snow accumulated when it is cooler, so we can use ice cores to track temperature changes through time, just as we can use calcium carbonate nodules or foraminifera shells.

We can also look at windblown particles trapped in the ice. Dirty ice is a telltale sign of dry, windy conditions, as salt and dust are blown in from the oceans and the continents beyond them. Then there are the bubbles—paleoclimatologists can measure carbon dioxide and methane levels in the air trapped within them.[14] Before the Industrial Revolution, the amount of carbon dioxide in the air was controlled largely by the world's oceans, with higher levels at warmer times. Methane came from bacteria in wetlands. Warmer and wetter conditions meant vegetation to match, and higher methane levels in the atmosphere.

Paleoclimatologists from the two ice core projects measured these climate proxies layer by layer: the water isotopes, dust and sea salt, carbon dioxide and methane levels. They built a superb archive of climate change in the past. As they worked backward through time into the last ice age, the cores became noticeably darker and dirtier—just as you'd expect given cooler and drier conditions. But there was much more detail than this. The cores taught them that climate was a fickle beast. It could stay stable for millennia, or it could turn on a dime. It seemed that conditions occasionally reached a tipping point when the tilt of the Earth and the distance from the Sun were just right, when the continents collided or separated, or when the ice sheets advanced or collapsed to the point that they changed patterns of ocean circulation. As Richard Alley has noted, climate flickered with alternating cold and warm decades or years, especially before big, long-term changes. Think of a lightbulb just before it dies. Offerings on the biospheric buffet must have flown on and off the table in a frenzy, and that had to have affected the lives of the people that lived through those climate changes.

Let's have a look at the details, starting with the last glacial maximum and working forward into the Holocene.[15] The northern hemisphere was dry and cold around 25,000 BP, brutally so in central Greenland. There was less carbon dioxide in the atmosphere then, and methane was at its lowest concentration in more than 100 millennia. The ice core layers from the time are darkened with dust and chock-full of sea salt; much of the northern hemisphere was beset by intense, recurring winter storms. The Earth was deep in the midst of an ice age that would last for more than 10,000 years. But there were occasional warm spikes. In fact, one of them seems to have occurred at 23,400 BP, just about the time Ohalo II was occupied.

The core changes abruptly at 14,700 BP, with the onset of what's called the Bølling-Allerød interval. Gas bubbles in the ice have higher concentrations of methane and carbon dioxide, indicating that levels of these atmospheric greenhouse gases rebounded. Oxygen isotope ratios also suggest that temperatures rose. There's less dust in the core at that point, too—another indication of the new calm, warm, and wet conditions. The world thawed. And while there was a brief cold blip within the interval, the Older Dryas stadial that separated Bølling from Allerød interstadials,[16] the northern hemisphere enjoyed mild temperatures on the whole for the better part of 1800 years. Desert and steppe gave way to wetland and forest, at least for a time.

But the Pleistocene wasn't over. The Big Freeze came around 12,800 BP. Paleoclimatologists call it the Younger Dryas stadial, a last gasp of the great ice age. Oxygen isotope ratios in the ice indicate that temperature plummeted in a few decade-long steps. The Greenland summit was, at its low point, nearly 30 degrees colder than it is today. The snow layers at the time were especially thin, a sign that drier conditions had returned. Heavy dust and sea salt accumulations point to an average wind strength double that of today. Atmospheric methane levels dropped as tropical wetlands dried and high-latitude ones froze. The ice age was back in full swing, and it continued unabated for 1200 years. Fortunately for us, the Earth hasn't experienced climate conditions like those since.

The Younger Dryas finally ended around 11,600 BP. It ended even more abruptly than it had started. The ice cores record the change in three steps of around five years each, spread over four decades. Most of the change came during the middle step. Snow accumulation doubled in three years as central Greenland warmed. Dust levels dropped when the winds abated, and the methane concentration rose as wetlands returned. Climate changed not over millennia or centuries, but years. It happened fast enough that anyone alive at the time would have realized it. The Pleistocene was finally over, and the conditions that replaced it quickly became warmer and wetter than any time since.

Matching the Climate and Archaeological Records

Greenland's ice sheet chronicles sudden, drastic climate swings that changed the world for those who lived through them. Imagine witnessing a drought that never ended, or seeing the arid steppe you roamed in youth overtaken by trees in your later years. Think of the "back in the day" stories that elders must have told around the hearth, late into the evening.

The first to settle at Abu Hureyra found a land of plenty in the warm, wet conditions of the Allerød interstadial. The Natufians there hunted animals and gathered plants in the marshes and gallery forests along the Euphrates, as well as in the woodlands beyond them. Nature provided ample food to support a small community without the need for villagers to move their homes or to grow their own. So they settled, and there was enough to sustain a steady population for 100 generations. But then nature pulled the rug out from under them. The cooler, drier, windier Younger Dryas winnowed out denser stands of water-dependent trees and bushes, and steppe returned to the middle Euphrates. Unrelenting conditions forced residents to rely more and more on drought-tolerant foods. This is just when, according to Moore, families started to cultivate gardens to supplement their dwindling supply of wild foods. At first they grew rye, and later added wheat and lentils in their fields.

The settlement at Abu Hureyra remained small and stable through the Allerød and into the Younger Dryas. There was no push to plant crops to feed an ever-growing population, nor did nature pull them into farming with a promise of the mild temperatures and steady rains that ensure a successful harvest. It seems that cultivation gave the village just enough to replace the foods that a return to cooler, drier times took from the biospheric buffet.

That doesn't mean, of course, that everyone responded to the Younger Dryas this way. While wild wheat and barley were staples for hunter-gatherers across the Fertile Crescent even in Ohalo II times, there's little evidence beyond Abu Hureyra that people began cultivating them before the Holocene. In some places, nature kept the biospheric buffet stocked well enough, and people weren't much affected.[17] Why put in the effort to plant when there's food all around you? But, in other places, the long dry spell had a profound effect. Hamlets were abandoned and their inhabitants scattered and spread as stands of fruiting trees and wild cereals shrank and vanished. Abu Hureyra, it seems, lay somewhere in the middle.

Scientists tend to prefer the consistent and predictable—where one action, like the onset of the Younger Dryas, causes a single, specific reaction, like cultivation. But life is complicated. People react in different ways to a given set of circumstances. Even if they did react in the same way, circumstances differed across the Fertile Crescent. Varying landscapes, from the Mediterranean coastal plain to the Jordan Valley to the hilly flanks of the Taurus and Zagros mountains, meant different spreads on the biospheric buffet in different places. Different mindsets, challenges, and opportunities meant different responses to climate change. For the people of Abu Hureyra, it meant taking control of their world.

The Neolithic didn't really kick into high gear across the Fertile Crescent until around 11,500 BP. Warmer and wetter conditions must have given people confidence that the effort of planting would provide a reliable return. An increase in atmospheric carbon dioxide would have also meant a larger yield. Larger permanent settlements appear and, within them, granaries that hint at intensive harvesting and surplus. Rather than a revolution, though, it looks as if the onset of agriculture may have been more of an evolution. Take the foothills of the Zagros Mountains in Iran, for example. Inhabitants of the village at Chogha Golan started cultivating wild cereal grains, lentils, and peas around the onset of the Holocene, if not before. But domesticated forms didn't become commonplace for another 1000 years.

That's just about when Abu Hureyra had its massive and sudden increase in, or influx of, people. Did natural growth of the population force villagers to ramp up production, ultimately providing the food needed for the burgeoning town? Or did a separate community of farmers move in from elsewhere and take over? The evidence during the key time period between the two well-sampled occupations is too limited to say for sure. Perhaps the answer is still buried in the mound, submerged under Lake Assad. Either way, favorable conditions had returned, and the Neolithic was well under way. The cultivated fields fed thousands, and civilization was just around the corner.

EFFECTS ON HUMAN BIOLOGY

We can see the shift from foraging to farming in the archaeological record, and understand it in the context of climate change revealed by Greenland's ice cores. But what about the people who lived through these changes? What effects did these things have on them and their descendants? If climate change and food choice drove human evolution, they must have had some impact on human biology in the more recent past too.

In some ways, of course, they haven't. We're no more "human" than those who came before the first farmers or herders. Look at the few remaining peoples whose ancestors never jumped on the Neolithic bandwagon. Today's hunter-gatherers are fundamentally no different from the rest of us so far as our humanity goes. Why would they be? We're only talking a few thousand years since people started planting and raising animals. While millennia might sound like a long time when your lifespan is less than a century, it is a blink of the eye for those of us who work with time scales in the millions of years. On the other hand, food production *has* affected our biology, albeit in subtle ways. By planting seeds and raising animals, Neolithic peoples began to transform themselves. We can see it in their bones and teeth.

Bioarchaeological Foodprints

In chapter 5 we learned that Nikolaas van der Merwe used carbon isotopes to trace the transition from foraging to farming in New York State during prehistoric times. Maize is essentially a tropical grass, with more heavy carbon (^{13}C) than is found in wild plants that far north. So early farmers who ate maize had relatively more heavy carbon in their bones and teeth than did the hunter-gatherers who came before them. But what about the Fertile Crescent? Can we use carbon isotopes to document the Neolithic transition there? Unfortunately we cannot. The two main crops, wheat and barley, both have similar proportions of ^{13}C to ^{12}C to the wild plants that hunter-gatherers would have eaten. The transition there is basically invisible to carbon isotope studies.

But the dental microwear is a different story. Bioarchaeologist Patrick Mahoney studied teeth from collections at Tel Aviv University—nearly

100 individuals from 11 sites ranging from late Pleistocene hunter-gatherers to Neolithic and Copper Age agriculturalists. The microwear shows an interesting pattern, with textures swinging back and forth over time from surfaces dominated by small pits and narrow scratches to ones with larger pits and broader scratches.[18] As Mahoney points out, these changes may reflect how foods were prepared as much as what people ate. Differences between the Ohalo II specimens and the Natufians surely reflect differences in diet. But later changes seem to relate to grit introduced from grinding stones, or to softening food by cooking in pots. Still, parsing these signals can be difficult, and much remains to be done to work out the details. If nothing else, though, Mahoney's study makes it clear that microwear does track differences over time in what people ate and how they prepared their foods.

Another bioarchaeologist, Theya Molleson from the British Museum, looked at the microwear on the teeth of Abu Hureyra residents back in the 1990s. Unfortunately, Moore and his team had found precious few skeletons in the all-important lower occupation levels. The trench seems to have missed the cemetery, if there is one. So Molleson couldn't look for changes between the Bølling-Allerød and Younger Dryas, or across the Pleistocene-Holocene boundary. On the other hand, she was able to compare those from the Natufian village as a whole with those from the Neolithic town that rose up later. And she found clear differences. Neolithic occupants had more microwear pits to show for their greater commitment to agriculture. They also had much more tooth wear on the whole than did earlier Natufians, likely owing to a greater dependence on grains that were husked and ground with gritty milling stones. Another big change came a couple of thousand years into the Neolithic occupation, once people started cooking in pots. They had less tooth wear and smaller microwear pits, which make sense given a diet of porridge or other moist, softened foods. Again, microwear can help document changes not just in diet, but also in the way foods are collected and prepared.

There's another category of foodprints that we haven't considered yet—paleopathology. Paleoanthropologists don't focus much on malnutrition, disease, or other identifiable conditions of the skeleton and teeth related to diet. But many bioarchaeologists do—they've got a lot more evidence to work with, and a massive baseline of forensics cases

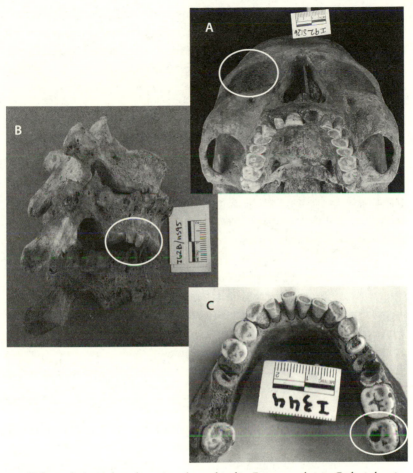

7.3. Paleopathological markers. A. cribra orbitalia; B. osteoarthritis; C. dental caries. All examples stem from the ancient city of Amarna in Egypt. Images courtesy of Jerry Rose.

today to compare it with. What did the shift from foraging to farming do to the human body? Ask a bioarchaeologist, particularly one that specializes in paleopathology.

The English philosopher Thomas Hobbes wrote in *Leviathan* (1668) that "the natural condition of mankind" was "solitary, poor, nasty, brutish, and short."[19] Surely the Neolithic Revolution must have improved the health and welfare of those who could settle down and enjoy the benefits of a reliable food supply. Early farmers had a full plate at dinnertime and were reassured by the knowledge that they were no longer bound by nature's whims. Human populations flourished. Look at Abu Hureyra. Renowned demographer Ansley Coale estimated that there were only eight million souls on the planet when the last ice age ended. Food production has allowed our numbers to swell a thousand-fold since. Thanks to the Neolithic Revolution, everything was, to quote singer Ethel Merman, "coming up roses."

But roses have thorns. When Molleson looked closely at the skele-tons from Abu Hureyra, for example, she found telltale signs that Neolithic women spent countless hours kneeling and hunched over grinding-stone slabs, milling grains. She found deformed toes, curved thigh bones, and arthritic knees and backs. Agriculture gave the towns-people food to support thousands, but it was hard work and must have made for a difficult life, especially for the women. Skeletons of the Natufians that came before them didn't have those pathologies; they surely had an easier time earning a living. This might at first glance seem surprising, but it wouldn't if you had attended the Man the Hunter symposium described in chapter 6. As early as the 1960s, cultural anthro-pologists began to realize that foragers actually spend less time working than do most of the rest of us. Some even called hunter-gatherer culture "the original affluent society."[20]

Bioarchaeologists like Molleson have added fuel to the fire. It wasn't just that the diets of early Neolithic farmers required harder work, they were often less healthful too. Agriculture may have given our ances-tors plenty of food, but it also meant a focus on a few easy-to-grow grasses that offer a poor nutritional base. In other words, early farmers may have had enough calories, but not the combination of nutrients their bodies needed. That's the difference between undernutrition and malnutrition.

The classic example comes from the New World. Maize provides plenty of energy, but it lacks some essential amino acids, vitamins, and minerals. You can read the nutritional deficiencies of early farmers in their short stature, and the distinctive anemia-induced spongy appear-ance of their skull bones. These are the go-to examples used by paleopa-thologists for decades to argue that the Neolithic Revolution brought with it malnutrition.[21] It might at first seem counterintuitive that more food led to malnutrition. But, in some cases, that's just what happened.

It's kind of like suffering dehydration while adrift in a life raft on the ocean, surrounded by water.

Bioarchaeologists have also looked to oral health as a window on overall well-being. As your teeth and gums go, so goes the rest of your body. Whether cause or effect, poor oral health has been linked to all sorts of diseases and conditions, from endocarditis and cardiovascular disease to premature birth and low birth weight, diabetes, HIV/AIDS, osteoporosis, Alzheimer's, and others.[22] And because teeth tend to preserve better than any other part of the body, there are plenty to work with from archaeological sites. Paleopathologists love dental remains as much as paleontologists do, and there's a huge literature on caries and periodontal disease in ancient peoples to prove it.

Much of this work has focused on the Neolithic transition in the Americas. We can start with the fact that carbohydrates feed the bacteria in dental plaque that cause cavities. These bacteria make lactic acid as a by-product when they're fed sugar and starch. More carbs, more lactic acid. More acid, more holes in your teeth. It should come as no surprise, then, that bioarchaeologists have found a fivefold increase in the average caries rate in the New World with the Neolithic Revolution, or that cavities are used as a foodprint to mark the change in diet from wild foods to maize.

Periodontal disease can be used as a foodprint in the same way. Plaque bacteria release toxins that cause the body to produce infection-fighting molecules called cytokines. The more plaque bacteria, the more the immune system responds. As the battle ensues, the gums, ligaments that anchor teeth to jaw, and jaw bones themselves come under friendly fire. You can think of the resulting tooth loss as collateral damage. Truth be told, the relationship between periodontal disease and agriculture is not quite as clear the one between caries and maize cultivation. It can be difficult to distinguish effects of disease from damage caused during or after burial, and the earliest stages may not even affect the jaw bone. Still, tooth loss did become more common in some places with the onset of the Neolithic, and some researchers have used this as a marker for the transition.

The oral health story seems pretty straightforward for the Americas. Things went downhill fast once maize entered the picture.[23] It's not nearly so clear for the onset of agriculture in the Fertile Crescent.

Israel Hershkovitz and his colleagues at Tel Aviv University compared the teeth of nearly 2000 Natufian and early Neolithic skeletons to document changes in oral health, but they found no marked differences in either caries rate or tooth loss. Could the Neolithic transition have meant less of a change in diet in the Near East than it did in the New World? Remember that wild barley and wheat were staples at Ohalo II and Abu Hureyra long before people in the Levant started domesticating cereal crops.

Nevertheless, Hershkovitz and his colleagues did find some differences between populations. Natufians had more periodontal disease and wore their teeth more quickly, and Neolithic peoples had more tartar buildup. But these may have related more to changes in how foods were prepared than what was being eaten. It's also telling that the team in Tel Aviv didn't find skeletal markers for differences in nutritional diseases, like anemia-related changes in skull bones. There were changes later, but not during the initial transition from foraging to food production.

Evolutionary Responses to the Neolithic Revolution

Skeletal and dental pathologies may show responses of the human body to changes in diet or how food is acquired and processed. They don't teach us about how the body has evolved to adapt to those changes. If anything, the pathologies suggest that Neolithic peoples *hadn't* adapted to an agricultural lifestyle all that well. But that's not surprising. A thousand years is only a few dozen generations. What kinds of evolutionary changes can we expect in that short time frame? They'd likely be subtle, like the enzymes that drive digestion.

Our ancestors had specific tools in their genetic shed for the job of breaking down the foods they evolved to eat. Different foods require different enzymes to cleave them into pieces small enough for our bodies to use. In other words, a change in diet might not make sense without genetic changes to make the enzymes we need to take full advantage of new foods. This is just the sort of thing we'd expect to evolve quickly, perhaps even on a time scale as short as that since the Neolithic Revolution. And it has. There are a growing number of examples that biological anthropologists use in their classrooms to illustrate recent human evolution.

We, on the whole, are better at digesting starch than are chimpanzees. It comes down to the number of copies of a gene called *AMY1*, which tells cells in your salivary glands to make the enzyme amylase. The more copies of *AMY1* you have, the more amylase in your saliva. Amylase breaks down starch, a process that begins in the mouth and continues in the gut. It acts like a cleaver to chop starch molecules into malt sugar, which in turn is broken down by another enzyme into the glucose we use to fuel our bodies. And humans have more *AMY1* copies than do chimpanzees. Perhaps more important, those of us with an agricultural heritage tend to have more *AMY1* copies than do hunter-gatherers. In other words, those of us descended from farmers have evolved over a very short time for more efficient digestion of cereal grains like wheat, barley, and rice, or root vegetables like potatoes and yams.

Lactase persistence is another example of recent evolution. Some of us are better at digesting milk than others. If you're lactose intolerant, a few licks of an ice cream cone might be enough to set off painful stomach cramps and other unpleasant effects. It's easy to understand how this could be a problem when it comes to reproduction. Still, it should come as no surprise that many adults can't digest milk. Most mammals shut down production of lactase, the enzyme that cleaves lactose into simple sugars, after weaning. Why waste the energy making something you're not going to use any more? But it's not a waste if you're a pastoralist and rely on the milk of goats, camels, or cows for nutrients. This has selected for mutations that keep lactase production going into adulthood, ones that have survived and spread since people started domesticating animals.

Enzymes give us great examples of how humans have evolved to their changing diets. They make it clear that the Neolithic transition can be used as a yardstick to gauge fine-scale evolution in our species. But they're not of much help beyond that for those of us who work with million-year-old fossils. We get more excited about adaptive changes in teeth and jaws. Has there been enough time since the transition from foraging to farming to see any? Absolutely. First and foremost, they get smaller, at least in some populations. Recall from chapter 6 that Louis Leakey and his colleagues thought hominins had evolved smaller teeth and wimpier jaws when tools started taking over their roles in food acquisition and processing. In fact, Leakey and his colleagues used tooth size and jaw robustness as defining traits for our biological genus, *Homo*. They argued that these traits mark a milepost on the road to humanity, a new dependence on culture and technology to make our way in the world. Cheek teeth also seem to get smaller from one *Homo* species to the next through the Pleistocene. The pattern is messy, but the trend is convincing.

The same holds true within our species, even in the short time frame of the Holocene. Neolithic peoples have smaller teeth than their hunter-gatherer predecessors in parts of Africa, Europe, and Asia. Even in the Levant, where we don't see a change in caries rate or tooth loss, early farmers had narrower cheek teeth than their Natufian ancestors. The explanation that Leakey and his colleagues proposed for the genus, that technological innovations and a change in diet meant less selective pressure to maintain large chewing platforms, has also been applied to early farmers. In this case, milling grains and, later, cooking food in ceramic pots led to softer diets. You don't need big teeth and powerful jaws to chew gruel or porridge.

But this doesn't explain why teeth got smaller. Relaxing selective pressure might do it if nature followed a "use it or lose it" model, but it doesn't. There must have been an advantage to smaller teeth, or at least a disadvantage to larger ones. The most reasonable explanation to me is one proposed years ago by bioarchaeologists James Calcagno and Kathleen Gibson at Loyola University in Chicago. They noticed that Neolithic people from northern Sudan had both smaller jaws and smaller teeth than their hunter-gatherer predecessors, and proposed a connection between jaw size and tooth size.

Think about it. We know that jaws grow longer with heavy use because of the way bone responds to strain.[24] Teeth, on the other hand, don't grow once they're formed. So nature has to guesstimate chewing stress, and resulting adult jaw length, to know how big to make teeth so that they'll fit right. In other words, teeth evolve to match the size of a jaw that develops within a specific strain environment. If the strain is less than nature expects, the jaw won't grow long enough to fit all the teeth. Those in the front might be crowded together or pushed forward, and those in the back might be impacted or emerge tilted or twisted. Dental crowding can lead to caries and malocclusion, both of which would select against larger teeth. Impacted wisdom teeth have their own set of problems, at least according to most dentists in the United States. Long story short, a shift from tubers and gazelle jerky to porridge and gruel should reduce chewing stress and jaw size, leading to selection for smaller teeth. We'll get into more details about this in chapter 8.

As an interesting sidenote, peoples in the Levant bucked the global trend in tooth and jaw length reduction during the shift from foraging to farming, much as they did for caries rate. On the other hand, early Neolithic peoples in the Levant did have narrower teeth, tongue to cheek, than the Natufians who preceded them.[25] Some have explained this as a way of keeping caries rates down by balancing a more cariogenic diet with less surface for cavities to take hold. A lot of work remains for us to understand the details of why teeth have shrunk as they have over the past few thousand years, but it's clear that the Neolithic Revolution had something to do with it.

FINAL THOUGHTS

The central premise of this book is that climate change drove human evolution, in large part by swapping out food options available on the biospheric buffet. But let's return to the question posed at the beginning of this chapter: What has it done for us lately? The answer is—a lot, especially in the past dozen millennia. I can accept that human ambition and changing views of our place in nature played a role in the transition from foraging to farming. Meeting the needs of a growing population must have too. But could the Neolithic have happened without the right climate conditions? It is hard to imagine that the co-occurrences of first cultivation at Abu Hureyra with the onset of the Younger Dryas and spread of agriculture across the Fertile Crescent with the start of the Holocene are coincidental. Other animals modify their diets when changing conditions require it. Humans must have too. The big difference in our case is that we began restocking the buffet ourselves, by planting seeds and raising animals. Raphael Pumpelly got a lot wrong, but he was right about one thing. Climate change at the end of the last great ice age touched off agriculture, at least in some places, and in doing so set in motion a cascade of events that led us to where we are today.

CHAPTER 8

Victims of Our Own Success

We're hardly the most successful species on the planet. By the numbers, there are less than eight billion of us, but there are millions of billions of ants. In terms of biomass, there's a good 50% more cow than human. If you measure success by species longevity, we've been here a couple of hundred thousand years, whereas the horseshoe crab *Limulus* has remained nearly unchanged for 150 million years.[1] What about our impact on the world? The total carbon we've pumped into the atmosphere doesn't hold a candle to the amount of oxygen put there by cyanobacteria 2.3 billion years ago. We've scarcely scratched the surface in light of the Great Oxygenation Event that fundamentally changed the history of life on our planet. We're so self-important. As comedian George Carlin said in his HBO concert, *Jammin' in New York*, "the planet has been through a lot worse than us," and will undoubtedly eventually "shake us off like a bad case of fleas."[2] Dominion over all living things, indeed.

One place we have done well, though, is the spread of our species. We've settled everywhere from Arctic Greenland to sub-Antarctic Navarino Island; from the parched, barren Atacama Desert to the rain-drenched cloud forests of Meghalaya; from the high Andes to the shores of the Dead Sea. The only other mammal that comes close to our distribution is the brown rat, and that's only because it has tagged along with us to most of the places we have gone. If for nothing else, we can pat ourselves on the back for the variety of ecosystems in which we've managed to eke out a living. That's our point of pride, the one our parent species would post on its refrigerator if it had one.

We built the case in chapter 6 that our ever-changing world winnowed out the pickier eaters among us. Nature has made us a versatile species, which is why we can find something to satiate us on nearly all of its myriad biospheric buffet tables. It's why we have been able to change the game, transition from forager to farmer, and really begin to consume our planet. But it has also meant that we can fuel our bodies with some pretty egregious things that we never evolved to eat—fried butter, hot dogs, cotton candy. It becomes clear, as we grapple with the difference between what we *can* eat and what we should, that we've become victims of our own success.

What is the natural human diet? This final chapter leads us directly into one of the most enduring debates in human history. Are we intended to eat meat, or are we innately vegetarian? Expressed in its modern form, should we cut saturated fats or carbs? Here we take a fresh look at this age-old question. From our vantage point, there doesn't seem to be an answer because the question itself is built on a fundamental misunderstanding of the evolution of human diet. Still, that does not mean we can't benefit from asking it. This book ends where it started, with our teeth and what they can teach us about ourselves.

FROM PYTHAGORAS TO PALIN

People have been debating the natural human diet for thousands of years, often framed as a question of the morality of eating other animals. The lion has no choice, but we do. Take the ancient Greek philosopher, Pythagoras, for example. He is best known to middle-school math students for $a^2 + b^2 = c^2$, but to ethical vegetarians he's a patron saint. "Oh, how wrong it is for flesh to be made from flesh; for a greedy body to fatten by swallowing another body; for one creature to live by the death of another creature!"[3] The argument hasn't changed much in 2500 years. Have a look at PETA's website.[4] But today we also have Sarah Palin. As she wrote in *Going Rogue: An American Life*, "If God had not intended for us to eat animals, how come He made them out of meat?"[5] I hate to say it, but she has a point. While according to Genesis 1:29, Adam and Eve may have been vegetarians, God was pretty explicit about giving meat to Noah in Genesis 9:3.

So what *were* we "designed" to eat? We're clearly not carnivores. There's a great story about a prank students tried to pull on naturalist Georges Cuvier in the early nineteenth century. One woke him up in the middle of the night raving, "Cuvier, I have come to eat you!"

He was wearing a devil's costume. Cuvier opened his eyes, took one look at the prankster, and without missing a beat replied, "all creatures with horns and hooves are herbivores." It's said he then rolled over and fell back to sleep.[6] We don't have horns or hooves, but we also don't have the teeth or claws of a mammal evolved to kill and eat other animals.[7]

That doesn't mean our distant ancestors were vegetarians, though. As we learned in chapter 6, early *Homo* invented weapons and cutting tools in lieu of sharp carnivore-like teeth. It's much easier to strike a flake off a cobble than to evolve bladed molars. There's no explanation other than meat eating for the fossil animal bones riddled with stone-tool cut marks at places like Olduvai Gorge. Creating tools also gave hominins more food options than specialized teeth would allow. Besides, early *Homo* did actually have somewhat sharper teeth, and perhaps also smaller, simpler guts than its *Australopithecus* forebears (recall our visit with Leslie Aiello in chapter 6). Just because we don't look like cats or dogs doesn't mean we aren't "supposed" to eat meat. As for the ethical dilemma of eating something with a face, that's an entirely personal decision. But it's not one that should be made based on the argument that meat consumption is somehow "unnatural."

THE CEREAL KILLERS

Gluten isn't unnatural either. Despite the pervasive call to cut carbs, there's plenty of evidence that cereal grains were staples, at least for some, long before domestication. In chapter 7 we learned that people at Ohalo II ate wheat and barley during the peak of the last ice age, more than 10,000 years before these grasses were domesticated. Paleobotanist Amanda Henry and her colleagues have even found starch granules trapped in tartar on Neandertal teeth dating back 40,000 years.[8] These had the distinctive shapes of barley and other grains, and the telltale damage that comes from cooking. There's nothing new about cereal consumption.

This leads us to the Paleolithic Diet. I'm often asked for my thoughts about it. Sometimes I think about Woody Allen's character, health-food-store owner Miles Monroe, in the movie *Sleeper*.[9] Monroe was cryogenically frozen in 1973, and thawed 200 years later by a pair of lab-coat-clad scientists, Doctors Agon and Melik. Their conversation went something like this:

> Agon: Has he asked for anything special?
> Melik: Yes, this morning for breakfast he requested something called *wheat germ*, organic honey and tiger's milk.
> Agon: [Laughs] Oh, yes. Those were the charmed substances that some years ago were felt to contain life-preserving properties.
> Melik: You mean there was no deep fat? No steak or cream pies or hot fudge?
> Agon: Those were thought to be unhealthy, precisely the opposite of what we now know to be true.
> Melik: Incredible.

I'm not really a fan of Paleolithic diets. I like pizza and French fries and ice cream too much. Most of my colleagues feel much the same way. I remember having dinner with several of them after a workshop on the evolution of human diet many years ago, and ordering mozzarella sticks and potato skin appetizers for the table. I also recall a twinge of guilt because Boyd Eaton was with us. Eaton was lead author of a seminal paper published in the *New England Journal of Medicine* and a follow-up book called the *Paleolithic Prescription*.[10] Eaton is, in the words of Paleo Diet guru Loren Cordain, "the Godfather of the Paleo Movement."[11] He and his colleagues built a strong case for discordance between what we eat today and what our ancestors evolved to eat.[12] The mismatch, they argued, accounts for most of the chronic degenerative disease that plagues our health care systems. We are, in effect, "Stone Agers in the fast lane," with our diets changing too quickly for our genes to keep up. The result is heart disease, diabetes, cancer, and many other ailments. Metabolic syndrome, a cluster of conditions including elevated blood pressure, high blood-sugar level, obesity, and abnormal cholesterol levels, is pervasive in our fast-food world. If only we could return to the diets of our ancestors . . .

It's a compelling argument. Our bodies didn't evolve for a food environment of bread and milk, peanuts and beans, corn syrup and table sugar, cola and coffee. Think about what might happen if you put diesel in an automobile built for regular gasoline. The wrong fuel can wreak havoc on the system, whether you're filling a car or stuffing your face. It just makes sense, so it's no surprise that Paleolithic diets are hugely popular right now. Not only is "Paleolithic" Google's top trending diet search, but many celebrities, from Miley Cyrus to Matthew McConaughey, have reportedly "gone Paleo."[13] There are lots of variants on the general theme, but foods rich in protein and omega-3 fatty acids show up again and again. Grass-fed cow meat and fish are good, and carbohydrates should come from nonstarchy fresh fruits and vegetables. On the other hand, cereal grains, legumes, dairy, potatoes, and highly refined and processed foods are out. The idea is to eat like our Stone Age ancestors: you know, spinach salads with avocado, walnuts, diced turkey, and the like.

The menu sounds great, but what are its real nutritional costs and benefits? I'm not a dietitian and cannot speak to the issue with authority, but a panel of 20 specialists assembled by *US News & World Report* can and did.[14] They ranked the Paleo Diet 36th among 38 popular diets. It scored poorly in every category considered, averaging 2.3 out of five stars for weight loss, ease of following, nutrition, safety, diabetes, and heart health. The British Dietetic Association even called it "a sure-fire way to develop nutrient deficiencies, which can compromise health and your relationship with food."[15] Paleolithic diets continue to create a stir, and there are countless books, articles, and blogs peddling opinions on them, both for and against.[16]

What I take issue with is less the nutritional costs and benefits of Paleolithic diets than their evolutionary underpinnings. Any diet that drains fat reserves means not meeting daily caloric needs. I find it hard to believe that nature would select for us to focus on foods that don't provide the nutrients required to maintain the body. Yes, different people metabolize calories in different ways, but the goal is always the same, to balance energy consumed with that expended over the long term. And there are many ways to do that, which seems to me to be the take-home message in the clinical practice guidelines of the American College of Cardiology, American Heart Association, and Obesity Society.[17] This leads us directly into the heart of the matter.

The Paleolithic diet is a myth. Think about the biospheric buffet. We learned in chapter 2 that food choice is as much about what's available to be eaten as it is about what a species evolved to eat. Just as foods appear and disappear from the forest throughout the year, dishes have been put out and taken away as our world has changed over time. That change has driven our evolution. Even if we could (and we can't) reconstruct the glycemic load; fatty acid, macro-, and micronutrient composition; acid/base balance; sodium/potassium ratio; and fiber content of foods eaten by a specific hominin, the information would be meaningless for planning a menu based on our ancestral diet. Because our world was ever changing, so too was the diet of our ancestors. Focusing on an ancestor at one point in our evolution would be futile. We're a work in progress. What was the ancestral human diet? The question itself makes no sense.

Hominins were spread over space too, and those living in the forest by the river surely had a different diet from their cousins on the lakeshore or the open savanna. It's also impossible to plan a menu based on our ancestors' food choices because these surely varied in different places, even at a single moment in time. Consider some of the recent hunter-gatherers who have inspired Paleolithic diet enthusiasts. The Tikiġaġmiut of the north Alaskan coast lived almost entirely on the protein and fat of marine mammals and fish, while the Gwi San in Botswana's Central Kalahari took something like 70% of their calories from carbohydrate-rich, sugary melons and starchy roots. Human foragers managed to earn a living from the larger community of life that surrounded them in a remarkable variety of habitats, from near-polar latitudes to the tropics. As we've learned, few other mammalian species can make that claim, and there is little doubt that dietary versatility has been key to the success we've had. Recall our visit with Rick Potts to Olorgesailie in chapter 4. Increasing climate fluctuation through the Pleistocene sculpted our ancestors—whether their bodies, or wit, or both—for the dietary flexibility that's become a hallmark of humanity.[18]

Another point is that, as evolutionary biologist Marlene Zuk pointed out in her book *Paleofantasy*, we're actually doing pretty well in the fast lane.[19] The 12,000 years we've had to begin adjusting to "modern" lifestyles seems to be plenty of time. We're not just flexible, we're malleable and easily changed. As we learned in chapter 7, the more copies of the *AMY1* gene the more there is of the enzyme in saliva that breaks down starch. Humans average three times the number of *AMY1* genes that chimpanzees have, and descendants of farmers tend to have more than children of hunter-gatherers. *AMY1* has spread like wildfire in the 500 generations since the people at Abu Hureyra first planted barley. Then there are those variants of the gene that maintain production of the enzyme that breaks down milk sugar. More of us with pastoralist pedigrees can tolerate milk into adulthood.

My favorite example, though, takes us back to Greenland. Danish medical researchers Hans Olaf Bang, Jorn Dyerberg, and Aase Brøndum Nielsen published a study in the *Lancet* in 1971 suggesting that high levels of fat from fish and marine mammals protect native Inuit people on the island from heart disease.[20] This made quite a splash, and led directly to the omega-3 craze that's still with us today. Fish-oil supplements are a billion-dollar-a-year industry in the United States alone.

But the story has become much more complex and much more interesting in the past few years. Berkeley geneticist Rasmus Nielsen and his colleagues recently scanned the genomes of Greenland Inuit looking for genetic signatures of adaptations to the extreme conditions there.[21] They found mutations in a cluster of genes for enzymes that regulate fat, particularly omega-3 fatty acids. Nielsen and his colleagues suggest that these people keep their blood-lipid levels within a healthy range *despite*, not because of, their high fat intake. Greenlanders get more fat from their food, so their bodies have evolved to make less, to compensate. If Nielsen is right, a diet of marine mammals and fish might not be so heart-healthy if you're not Inuit.

So is fish oil the twenty-first-century snake oil? Should we shake our heads in wry amusement like Agon and Melik did in *Sleeper*? The original work by Bang and his colleagues has been challenged on a number of grounds, and clinical studies published in top medical journals have called into question the benefits of fish-oil supplements.[22] I'm sure we haven't heard the last word on the subject, and time will tell whether fish oil goes the way of wheat germ, organic honey, and tiger's milk, but Rasmus Nielsen's study gives us a lot to think about. The genes that limit fatty-acid production in Greenlanders provide even more evidence that dietary adaptations can evolve quickly, and for various types of food. They also underscore the fact that different people have different adaptations depending on where their ancestors lived and what they ate. And those diet differences can be dramatic. The take-home message is, again, we did not evolve under one specific set of circum-stances, so there is no single ancestral human diet for us to emulate.

OPEN YOUR MOUTH AND LOOK IN A MIRROR

While the question of what we evolved to eat makes no sense in light of our new understandings, that doesn't mean we cannot benefit from an evolutionary perspective on relationships between diet and health. Let's close out this book where it started, with teeth. Open your mouth and look in a mirror. Do you have any fillings or crowns? Are your lower front teeth crowded together, or do your uppers jut out in front of them? Have your wisdom teeth been pulled or cut out? Do you have gingivitis or gum disease? Nearly all of us answer "yes" to at least one of these questions.[23] Other animals get an occasional cavity or erupt a crooked tooth, but these things are relatively rare in nature. That begs us to think about our oral health from an evolutionary perspective. When we do, it comes to mind that the chemical, biotic, abrasive, and stress environments in our mouths today are unique. While there's little doubt that our teeth evolved under many different conditions as diets changed over time and space, we can be certain that at no point in our distant ancestry were they bathed in milkshake or coated in taffy. It's no wonder, then, that we find little evidence for dental decay, gum dis-ease, or orthodontic disorders in fossil hominins.

As we learned in chapter 7, carbohydrates feed the plaque bacteria that cause caries and periodontal disease. Those bacteria make the lactic acid that corrodes tooth surfaces. They also produce toxins that rouse our immune systems to action, causing collateral damage to our gums, connective tissue, and the bone that supports the teeth. The connec-tion between the Neolithic Revolution in some places, especially in the Americas where maize was domesticated, and a drop in oral health makes sense given increasing carbohydrate consumption. But it didn't hap-pen that way everywhere. Recall that there was no change in caries fre-quency when people in the Levant started planting wheat and barley, and that Natufian hunter-gatherers actually had *more* gum disease than the Neolithic peoples who followed them. Where caries rates did rise with the onset of agriculture, early farmers still averaged only a small fraction of the cavities we have today.

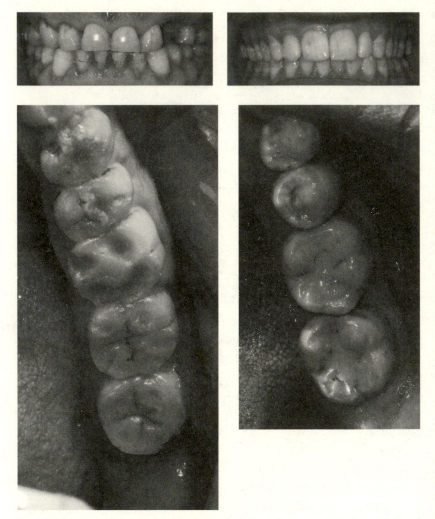

8.1. Comparison of the teeth of a modern hunter-gatherer consuming a traditional foraging diet (*left*) and an urban dweller consuming an "industrial-age" diet (*right*). Both live in northern Tanzania.

The real problem is sucrose, or table sugar, which helps the bacteria responsible for cavities stick to teeth and makes it easier for them to colonize, accumulate, and produce the lactic acid. Because table sugar didn't become widely available until the Industrial Revolution, a dozen generations ago at the most, there just hasn't been enough time for us to evolve a way to keep it from rotting our teeth.

So is there any way to use this information to help us manage our oral health today? Let's consider orthodontics. The first thing that struck me when I looked into the mouths of Hadza foragers (see chapter 6) was that they've got a lot of teeth. Most have a dozen working molars, whereas the typical dental patient in the United States has eight. Our wisdom teeth are impacted 10 times more frequently than are those of hunter-gatherers. They either never erupt, or dentists yank them out because there just isn't enough room for them in the back of the mouth. For that matter, there isn't enough room in the front of the mouth, either. Most of us have lower front teeth crowded together or crooked, and an overbite, where the uppers jut forward in front of the lowers. Early hominins and recent foragers typically had a tip-to-tip bite between upper and lower incisors, and the edges of their lowers were aligned to form a flawless arch. That's because the sizes of their teeth and jaws matched perfectly.

Recall from chapter 7 that the length of your jaw depends on the strain it has to bear during growth. Tooth size, on the other hand, is genetically programmed based on nature's best guess of ultimate jaw length. A harder or tougher diet can mean more than enough room for all the teeth, but a softer, mushier one may leave the jaw too short. Insufficient space can mean crowding and overbite at the front of the tooth row and impaction at the back. Perhaps, then, our orthodontic problems today relate to the fact that we eat mostly softened, processed foods. We just don't give our jaws the workout they need. Dental anthropologist Robert Corruccini has seen the difference in occlusion between urban dwellers and rural peoples in and around Chandigarh in India—soft breads and mashed lentils on the one hand, coarse millet and tough vegetables on the other. He also saw it from one generation to the next in the Pima of Arizona with the opening of a commercial food-processing facility on the reservation.[24] Diet makes a huge difference. I remember asking my wife not to cut my daughters' meat into such small pieces when they were young. "Let them chew," I begged. She replied that she'd rather pay for braces than have them choke. I lost that argument.

Nevertheless, crowded, crooked, misaligned, and impacted teeth are huge problems today. Not only are there aesthetics to worry about, but these problems can affect chewing, lead to increased decay, and compromise anchoring of teeth in the jaw. Nine out of ten of us have at least slight malocclusion, and about half could benefit from orthodontic treatment. Again, it's not that our teeth are too large, but that our jaws are too short to accommodate them. So why do most treatments of dental crowding and malocclusion involve pulling out or carving down teeth rather than growing longer jaws? One of my colleagues here at Arkansas, bioarchaeologist Jerry Rose, has joined forces with local orthodontist Richard Roblee with that very question in mind. Their recommendation? Clinicians should focus more on growing jaws, especially for children. For adults, surgical options for stimulating bone growth are gaining momentum too, and can lead to shorter treatment times.[25] We may be stuck with an oral environment our ancestors never had to contend with, but recognizing this can help us deal with it in better ways. Think about that the next time you smile and look in a mirror.

NOTES

INTRODUCTION

1. Woody Allen, *Love and Death*, directed by Woody Allen (Culver City, CA: MGM Studios, 1975), film.

CHAPTER 1: HOW TEETH WORK

1. The earliest mammals appeared around 225 million years ago. They are identified by the appearance of a new jaw joint capable of both vertical closing and horizontal movement for chewing.
2. See, for example, Henry Fairfield Osborn, *Impressions of Great Naturalists: Darwin, Wallace, Huxley, Leidy, Cope, Balfour, Roosevelt, and Others* (New York: Charles Scribner's Sons, 1928), 179. In his biographical sketch of Cope, Osborn reminisced, "Quaker by birth, he was a fighter by nature, both in theory and in fact. On one occasion, in the American Philosophical Society, a difference of opinion with his friend Persifor Frazer led to such a violent controversy that the two scientists retired to the hallway and came to blows! On the following morning I happened to meet Cope and could not help remarking on a blackened eye. 'Osborn,' he said, 'don't look at my eye. If you think my eye is black, you ought to see Frazer this morning!'"
3. Cope's father Alfred arranged for an honorary master's degree from Haverford's Board of Managers so Edward could teach there.
4. Thomas Williamson, Douglas Nichols, and Anne Weil, "Paleocene Palynomorph Assemblages from the Nacimiento Formation, San Juan Basin, New Mexico," *New Mexico Geology* 30, no. 1 (2008): 3–11.
5. Edwin Colbert, "Triassic Rocks and Fossils," in *New Mexico Geological Society Fall Field Conference Guidebook, 11: Rio Chama Country*, ed. Edward C. Beaumont and Charles B. Read (New Mexico Bureau of Geology and Mineral Resources, 1960), 55.

6. William King Gregory, "A Half Century of Trituberculy: The Cope-Osborn Theory of Dental Evolution with a Revised Summary of Molar Evolution from Fish to Man," *Proceedings of the American Philosophical Society* 73, no. 4 (1934): 180.

7. George Gaylord Simpson, "Mesozoic Mammalia, IV: The Multi-tuberculates as Living Animals," *American Journal of Science* 11, no. 63 (1926): 228.

8. J. William Schopf, *Cradle of Life: The Discovery of Earth's Earliest Fossils* (Princeton, NJ: Princeton University Press, 2001).

9. George Gaylord Simpson, "Paleobiology of Jurassic Mammals," Palaeobiologica 5 (1933): 155.

10. Christopher Beard, *The Hunt for the Dawn Monkey: Unearthing the Origins of Monkeys, Apes, and Humans* (Berkeley, CA: University of California Press, 2004).

11. This is an interesting case because insectivorous primates also have sharp teeth. On the other hand, there aren't any large primates that specialize on insects, so if we look at the *combination* of tooth size and shape, we can still work out relationships reasonably well. This becomes a case of, "all species with the *combination* of forms X, Y, and Z use them for function A."

12. Karen Hiiemae wrote in "Masticatory Function in the Mammals," *Journal of Dental Research* 46, no. 5 (1967): 883 that teeth are "an essentially passive element in the active masticatory apparatus and are dependent on the movements of the mandible for their functional interactions."

13. Peter Lucas's later book, *Dental Functional Morphology* (Cambridge: Cambridge University Press, 2007), grew out of this work. It tops the "must-read" list of academic works on the subject.

14. Georges Cuvier, *Essay on the Theory of the Earth*, trans. Robert Kerr (Edinburgh: William Blackwood, 1913).

15. This is all laid out in my book, *Teeth: A Very Short Introduction* (Oxford: Oxford University Press, 2014).

CHAPTER 2: HOW TEETH ARE USED

1. There are half a dozen types of nutrient that primates use for fuel and to supply the raw materials needed for everything the body is

and does: carbohydrates, proteins, lipids, vitamins, minerals, and water. Any of the first three can be burned for energy. There are some *essential* nutrients that primates cannot synthesize themselves and typically get through food. These include about half the amino acids needed to make our proteins, vitamins, minerals, and a couple of fatty acids. Individuals need to consider these carefully when choosing their diets. On the other hand, primates can build many of the nutrients they need from others, or absorb them from microbes living in their guts. This means flexibility in food choice.

2. Robert Sussman, "The Lure of Lemurs to an Anthropologist," in *Primate Ethnographies,* ed. Karen Strier (Boston, MA: Pearson Education, 2014), 34–45.

3. Alison Jolly, *Lemur Behavior: A Madagascar Field Study* (Chicago: University of Chicago Press, 1966).

4. Leaves are everywhere in the forest canopy. They contain simple sugar molecules strung together into long chains called polysaccharides, each of which can be hundreds or thousands of units long. The potential energy available to a primate is staggering. But it isn't easy to access. Polysaccharides must be hewn into simple sugars to pass from the gut to the bloodstream, and primates don't make the chemicals needed to break the bonds that hold them together. Leaf-eating primates have figured out a clever way around this, though. They fracture the leaves as best they can by chewing to increase exposed surface area, and then pass them on to microorganisms in the gut that can break down polysaccharides. But it's a slow-going, inefficient process. And it takes a lot of energy to maintain the large, complex digestive tracts needed to house enough gut microbes to do the job. Many folivorous primates move slowly and/or nap frequently to conserve energy. Fruit flesh, on the other hand, has simple sugars, and insects have proteins that can be easily digested and don't require big, costly guts. These foods offer a higher net energy yield, but they also have their own set of drawbacks. Fruits and insects are less dependable and patchily distributed in space and time. Access to fleshy fruits can also be limited by competition, and fruiting trees can attract predators ready to make a quick snack of a fruit eater. And when you consider the work it takes to find and catch bugs

and spiders, they may not deliver enough calories to make the effort worthwhile, especially for larger primates that need more fuel.

5. Richard Preston, *The Hot Zone: The Terrifying True Story of the Origins of the Ebola Virus* (New York: Anchor, 1995), 256.

6. George Schaller, *Year of the Gorilla* (Chicago: University of Chicago, 1964); Dian Fossey, *Gorillas in the Mist* (Wilmington, MA: Houghton Mifflin, 1983).

7. Richard Owen, "On the Gorilla (Troglodytes gorilla, Sav.)," *Proceedings of the Zoological Society of London* 27, no. 1 (1859): 19.

8. The San Francisco Zoo group is well known in the research community. In fact, the most famous gorilla of our time, Koko, comes from this very group. She was at first loaned by the zoo to Penny Patterson for a language study in 1971.

9. Phenologists typically use the terms *irradiance* and *photoperiod* when they discuss the effects of solar radiation on plant life-cycle events. Irradiance is a measure of radiative flux received by a surface, the amount of solar energy striking it. Photoperiod refers to day length, or the number of hours a plant receives light each day.

10. Plants and animals in an ecological community often act in step like the working parts of a watch. Can we be sure that the particular species in a forest coevolved phenological patterns and other traits because of interactions of their ancestors with one another? Some of the attributes that allow two species to coexist may have evolved long before their ancestors ever came into contact. I would argue, though, that for our purposes it doesn't matter. The point is that they do coexist, and that their interactions give an ecosystem its unique character.

CHAPTER 3: OUT OF THE GARDEN

1. Arthur C. Clarke, *2001: A Space Odyssey* (New York: New American Library, 1968), 24.

2. Readers can experience the tale themselves, in Raymond Dart's own words, in his book, *Adventures with the Missing Link* (New York: Harper, 1959).

3. Primate fossils had actually already been found in East Africa, at Olduvai Gorge in Tanzania, and Koru, near Lake Victoria, in Kenya. But there was no easy way to check back in those days. There were

even fossil primates from Taung itself in the collections at the National Museum in Cape Town, though their descriptions had not yet been published.

4. Researchers today prefer to use an ape model for the timing of tooth eruption in *Australopithecus*, and believe the Taung child died around age three or four, given the teeth in its mouth at the time.

5. Raymond Dart, "*Australopithecus africanus*: The Man-Ape of South Africa," *Nature* 115, no. 2884 (1925): 195.

6. Charles Darwin, *The Descent of Man, and Selection in Relation to Sex* (London: John Murray, 1871), 199.

7. "*Eoanthropus dawsoni*," better known as Piltdown man, was at the time thought to be a Pleistocene human ancestor, with a large, humanlike brain, but primitive, apelike jaw and teeth. The specimen suggested to the scientific establishment that an increase in brain size drove the evolution of other humanlike features, including the teeth and jaws. It turned out that the "find" was a carefully orchestrated hoax. The skull was that of a recent human and the lower jaw belonged to an orangutan. The bits and pieces had been stained to look old, and they were broken and filed down to mask the fact that they did not fit together. No one ever came forward to admit the hoax, which has been the subject of many popular whodunit articles and books.

8. Darwin, *The Descent of Man*, 140.

9. Raymond Dart, "Taungs and Its Significance," *Natural History* 26 (1926): 319.

10. The museum was renamed the Ditsong National Museum of Natural History in 2010.

11. David Meredith Seares Watson, "Robert Broom, 1866–1951," *Obituary Notices of Fellows of the Royal Society* 8, no. 21 (1952): 40.

12. Broom, his colleagues, and their successors have found hundreds of fossil hominin specimens at Sterkfontein since, including four partial skeletons.

13. Travis Pickering, *Rough and Tumble: Aggression, Hunting, and Human Evolution* (Berkeley, CA: University of California Press, 2013).

14. Raymond Dart, "The Predatory Transition from Ape to Man," *International Anthropological and Linguistic Review* 1, no. 4 (1953): 209.

15. Charles Kimberlin Brain, "Fifty Years of Fun with Fossils: Some Cave Taphonomy-Related Ideas and Concepts that Emerged Between 1953 and 2003," ed. Travis Rayne Pickering, Kathy Schick, and Nicholas Toth (Gosport, IN: Stone Age Institute, 2007), 4.

16. Dart, "The Predatory Transition": 214.

17. Details about Robinson's life can be found in Becky Sigmon's biographical account, "The Making of a Palaeoanthropologist: John T. Robinson," *Indian Journal of Physical Anthropology and Human Genetics* 26, no. 2 (2007): 179–341.

18. Dutch-born entomologist Antonie Janse sold his extensive collection of moths, which had been kept in his home, to the South African government in 1945. The collection was put in the care of the Transvaal Museum, but there was no space available at the museum to house it. Janse was appointed honorary curator of Heterocera (moths), funds were made available to purchase some new drawers and equipment, and the collection that had been previously housed at the museum was instead moved to Janse's home to join the newly accessioned specimens! According to Janse's obituary by Lajos Vári and Alexey Daikonoff, "Antoinie Johannes Theodorus Janse," *Journal of the Lepidopterists' Society* 25, no. 3 (1971): 211–13, an assistant was hired to "continue his work eventually."

19. Charles Kimberlin Brain, *The Transvaal Ape-Man-Bearing Cave Deposits*, Transvaal Museum Memoir no. 1 (Pretoria: Transvaal Museum, 1953).

20. Researchers now believe the sands accumulated later, in the late Pliocene or even the early Pleistocene.

21. John Robinson, "Adaptive Radiation in the Australopithecines and the Origin of Man," in *African Ecology and Human Evolution*, ed. F. Clark Howell and François Bourlière (Chicago: Aldine, 1963), 409.

22. Colin Groves and John Napier, "Dental Dimensions and Diet in Australopithecines," *Proceedings of the VIIIth International Congress on Anthropological and Ethnographic Sciences* 3 (1968): 274.

23. Clifford Jolly, "The Seed-Eaters: A New Model of Hominid Differentiation Based on a Baboon Analogy," *Man* 5, no. 1 (1970): 21.

24. Some at the time were starting to think that *Australopithecus* was ancestral to both *Paranthropus* and early *Homo*, and Jolly acknowl-

edged the possibility that the former was of "primitive (Phase I) stock." But, like his predecessors, Jolly believed them likely too late and too cerebrally advanced to be primitive.

25. Phillip Tobias, *Olduvai Gorge: The Cranium of* Australopithecus *(*Zinjanthropus*) boisei* (Cambridge: Cambridge University Press, 1967), 243.

26. We often abbreviate the first, or generic, names of species. *Australopithecus afarensis* becomes *Au. afarensis*, *Homo habilis* becomes *H. habilis*, etc.

27. Ellen Ruppel Shell, "Waves of Creation," *Discover Magazine* (May 1993): 55–61.

28. This idea has morphed into the Red Queen hypothesis, which we'll learn about in chapter 5. Red Queen stands in contrast to the Court Jester hypothesis, which builds upon Eldridge and Gould's punctuated equilibrium model. The basic argument comes down to whether abiotic forces (like climate) or biotic ones (competitors, predators, or prey) are the main drivers of evolution.

29. Elisabeth Vrba, "Evolution, Species and Fossils: How Does Life Evolve?" *South African Journal of Science* 76, no. 2 (1980): 82.

30. Genetic evidence now dates the last common ancestor of humans and chimpanzees to about 7–8 mya. Also, new finds and refined dating techniques indicate that hominins spread into Asia much earlier than was thought when Vrba developed her turnover-pulse hypothesis, likely before 1.8 mya.

CHAPTER 4: OUR CHANGING WORLD

1. The Earth system includes the atmosphere, hydrosphere, lithosphere, and biosphere. The challenge in Earth system science is to understand how these interdependent parts interact. Climate studies are an important part of the process.

2. Global warming is real. Some of the details can be debated, but there's very little controversy among scientists about the big picture. See the Royal Society and the US National Academy of Sciences, *Climate Change: Evidence and Causes* (2014), http://dels.nas.edu /resources/static-assets/exec-office-other/climate-change-full.pdf.

3. Peter deMenocal, personal communication (e-mail to author, December 2015).

4. Mark Maslin, *Climate: A Very Short Introduction* (Oxford: Oxford University Press, 2013).

5. Milutin Milanković, *Mathematische Klimalehre Und Astronomische Theorie Der Klimaschwankungen* (Berlin: Borntraeger, 1930).

6. See Peter deMenocal and Jessica Tierney's article in Nature Education's online Knowledge Project, "Green Sahara: African Humid Periods Paced by Earth's Orbital Changes," *Nature Education Knowledge* 3, no. 2 (2012): 12.

7. JOIDES stands for "Joint Oceanic Institutions for Deep Earth Sampling," a consortium of hundreds of scientists from around the world. What began in the late 1960s as a largely US-funded drilling program evolved into a major international collaborative effort—first the Ocean Drilling Program and then the Integrated Ocean Drilling Program, which continues today. The ship's namesake was the British Navy sloop *HMS Resolution*, which had explored the Pacific Ocean, its islands, and Antarctica back in the 1770s under the captainship of James Cooke.

8. Peter Clift, "Waiting for Core," *JOIDES Resolution: Science in Search of Earth's Secrets* (blog), March 8, 2014 (11:59 p.m.), http://joides resolution.org/node/3515.

9. Maslin, *Climate*.

10. Yves Coppens, "East Side Story: The Origin of Humankind," *Scientific American* 270, no. 5 (1994): 88–95.

11. Thure Cerling, personal communication (e-mail to author, October 2016).

12. Plants use light energy from the Sun to reorganize water and carbon dioxide into sugar and oxygen. The process is complex but, in the end, a simple hexose sugar molecule ($C_6H_{12}O_6$) and six dioxygen (O_2) molecules are produced from six molecules each of carbon dioxide (CO_2) and water (H_2O). The carbon in the CO_2 is converted, or "fixed," into the organic structure of a plant in one of three ways. The most common, used by trees, shrubs, and cool-season grasses, is called C_3 carbon fixation because a compound with three carbon atoms is produced during the process. Think of beans; fruits and vegetables; or wheat, barley, and oats. Other plants (sedges and

grasses at lower latitudes and elevations) use C_4 carbon fixation—they produce a compound with four carbon atoms instead. C_4 carbon fixation is more efficient in places with higher temperatures, more light, and less water available—which explains why C_4 grasses have spread across the African savannas. Important C_4 foods today include maize, sugarcane, and millet. Both C_3 and C_4 plants have a lower proportion of the heavier ^{13}C isotope than does atmospheric CO_2, but the C_3 process selects against it more than the C_4 process, so plants with different approaches to carbon fixation have different isotope ratios. The third type of carbon fixation, called CAM (or Crassulacean Acid Mechanism, for the plant in which it was first discovered), is used by cactuses, orchids, and some succulent plants. It results in an intermediate ratio of heavier to lighter carbon atoms.

13. We'll meet Glynn Isaac in chapter 6.
14. See Holli Riebeek, "Paleoclimatology: The Oxygen Balance," NASA Earth Observatory, last modified May 6, 2005, http://earth observatory.nasa.gov/Features/Paleoclimatology_OxygenBalance/.
15. And every 800,000 years after that.
16. There are great blogs from the field and descriptions of each project on the Smithsonian Institution's Human Origins Program website (http://humanorigins.si.edu/research/east-african-research-projects /olorgesailie-drilling-project) and the University of Arizona's Hominin Sites and Paleolakes Drilling Project website (https://hspdp .asu.edu/).
17. See Ann Gibbons's "How a Fickle Climate Made Us Human," *Science* 341, no. 6145 (2013): 474–79, and Elizabeth Pennisi's "Out of the Kenyan Mud, and Ancient Climate Record," *Science* 341, no. 6145 (2013): 476–79.

CHAPTER 5: FOODPRINTS

1. Quoted in Martin Meredith, *Born in Africa: The Quest for the Origins of Human Life* (London: Simon and Schuster, 2011).
2. Simpson, "Mesozoic Mammalia, IV": 228.
3. Alan Walker, Henrick Hoeck, and Linda Perez, "Microwear of Mammalian Teeth as an Indicator of Diet," *Science* 201, no. 4359 (1978): 908–10.

4. Wolpoff's theory also motivated Bob Sussman to study the lemurs of Madagascar, to understand how closely related primate species could coexist and divvy up limited resources (see chapter 2).

5. To give credit where it is due, Bernard Wood of George Washington University and Dave Strait, now of Washington University in Saint Louis, had suggested the year before that *Paranthropus* may have had a more flexible diet (eurytopic, as they called it) than most thought at the time. What do you feed a 400-pound gorilla? Anything it wants. What do you feed a *Paranthropus* with teeth the same size? You get the idea.

6. Leslie Poles Hartley, *The Go-Between* (London: Hamish Hamilton, 1953), 9.

7. Leslea Hlusko and I recently proposed ("The Evolutionary Path of Least Resistance," *Science* 353, no. 6294 (2016): 29–30) that functional morphology should follow the genetic path of least resistance. The probability of a given solution, even if suboptimal, to an adaptive problem should be inversely proportional to the evolutionary work, or number of mutations, required to get to it from a given starting point.

8. About one in a trillion carbon atoms is ^{14}C. These atoms form as a by-product of collision between nitrogen atoms and free neutrons in the atmosphere. You get ^{14}C (six protons, eight neutrons) from ^{14}N (seven of each) by swapping a neutron for a proton. Carbon 14, like other isotopes of carbon, combines with oxygen to form CO_2, which makes its way into plants during photosynthesis. While ^{13}C and ^{12}C are stable, ^{14}C is not, and decays back into nitrogen over time. Half of the original ^{14}C is gone in about 5730 years. Since the rate of decay is fairly constant, we can figure out when a sample of carbon formed by counting the number of remaining unstable atoms in a sample and comparing that to the number of stable ones.

9. Michael DeNiro and Sam Epstein, "Carbon Isotopic Evidence for Different Feeding Patterns in Two Hyrax Species Occupying the Same Habitat," *Science* 201, no. 4359 (1978): 906–8, and Walker, Hoeck, and Perez, "Microwear of Mammalian Teeth."

10. Nikolaas van der Merwe, Fidelis Masao, and Marion Bamford, "Isotopic Evidence for Contrasting Diets of Early Hominins *Homo habilis* and *Australopithecus boisei* of Tanzania," *South African Journal of Science* 104, no. 3–4 (2008): 153–55.

11. Tim White, et al., "*Ardipithecus ramidus* and the Paleobiology of Early Hominids," *Science* 326, no. 5949 (2009): 64–86. *Ardipithecus ramidus* was first discovered in the early 1990s by White and his team along the Awash River in Ethiopia's Afar Depression. The species dates to about 4.4 mya, older than *Australopithecus, Paranthropus,* or *Homo.* It ranks among the most primitive of the hominins, with a brain the size of a chimpanzee's, a thumb-like big toe, and thinner tooth enamel than later hominin species. That said, *Ardipithecus* also had smallish canine teeth, less of a muzzle than chimpanzees, and hints of evidence for some upright walking in the skeleton.

12. Quoted in Nicholas Rescher, *The Limits of Science* (Pittsburgh, PA: University of Pittsburgh Press, 1999), 13.

13. Lewis Carroll, *Through the Looking-Glass, and What Alice Found There* (London: Macmillan, 1882), 42.

14. John Harris, "Tribe Phacochoerini Warthogs," in *Mammals of Africa, Volume VI: Pigs, Hippopotamuses, Chevrotain, Giraffes, Deer and Bovids*, ed. Jonathan Kingdon and Michael Hoffman (London: Bloomsbury, 2013), 49.

CHAPTER 6: WHAT MADE US HUMAN

1. Richard Leakey, "Early Humans: Of Whom Do We Speak?" in *The First Humans: Origin and Early Evolution of the Genus* Homo, ed. Fred Grine, John Fleagle, and Richard Leakey (Berlin: Springer Science, 2009), 5.

2. Recall from chapter 3 that the larger-toothed species at Swartkrans was called *Paranthropus robustus* and the smaller one was named "*Telanthropus capensis.*" As an aside, by the time the new species was announced, Louis Leakey and his colleagues had put the nutcracker man species into the genus *Australopithecus.* It was later included with the species *robustus* in *Paranthropus.*

3. The name "*habilis*" was actually suggested by Tobias's mentor at the time, Raymond Dart. Dart had named *Australopithecus africanus* (see chapter 3) four decades earlier.

4. *Homo erectus* was well known from Asia at the time, but not from eastern or southern Africa.

5. When *erectus* was first found, Eugene Dubois named a new genus for it, "*Pithecanthropus.*"

6. Indeed, Sir Arthur Keith had argued many years before that 750 cm³ was a threshold of sorts for brain volume, a key milestone on the road to humanity. See the discussion on page 159 for more details.

7. Recent reanalysis suggests Jonny's child's skull may have been a bit larger than Tobias originally inferred—though still smaller than typical for *Homo erectus*.

8. Paleoanthropologists don't all agree on the bracketing dates for individual hominin species. Those from the Smithsonian's virtual hall of human origins, http://humanorigins.si.edu, seem to me as reasonable as any.

9. Also, Brian Villmoare of the University of Nevada, Las Vegas, and his colleagues recently found a jaw dating from 2.8 mya in the Afar region of Ethiopia that they attribute to early *Homo*. This could set the origin of our genus back another 400,000 years and throw a wrench into the works. See Brian Villmoare, et al., "Early *Homo* at 2.8 Ma from Ledi-Geraru, Afar, Ethiopia," *Science* 347, no. 6228 (2005): 1352–55. Time will tell.

10. Isaac would receive his PhD for his work at Olorgesailie (see chapter 4) eight years later.

11. Recall from chapter 3 that Maglio was documenting changes in elephant tooth size over time with the spread of grassland across East Africa.

12. Ecologists use the term *guild* for groups of species that play a similar role in a community of life. A large carnivore guild on an African savanna today might include lions, leopards, cheetahs, wild dogs, and hyenas.

13. Sonia Harmand from Stony Brook University and her colleagues have recently discovered earlier (3.3 mya) stone flakes purported to be tools in West Turkana, Kenya (Sonia Harmand, et al., "3.3-Million-Year-Old Stone Tools from Lomekwi 3, West Turkana, Kenya," *Nature* 521, no. 7552 (2015): 310–15). Also, Shannon McPherron, of the Max Planck Institute for Evolutionary Anthropology, and colleagues reported cut marks in bones dated to 3.4 mya in deposits in Ethiopia (Shannon McPherron, et al., "Evidence

for Stone-Tool-Assisted Consumption of Animal Tissues before 3.39 Million Years Ago at Dikika, Ethiopia," *Nature* 466, no. 7308 (2010): 857–60). If these interpretations and dates stand, they could lead to yet another rewrite for our definition of *Homo*.

14. Kim Hill and Hillard Kaplan interpret the data differently from their old mentor, Kristen Hawkes. They downplay the role of showing off, and emphasize taking care of the nuclear family as motivation for Aché men to hunt.

15. Richard Wrangham, *Catching Fire* (New York: Basic Books, 2009).

16. There's no question that our brains are big, but some animals have bigger ones. The sperm whale brain weighs more than five times ours. And if you consider relative rather than absolute size, the leaf-cutting ant's brain makes up 15% of its body mass, whereas ours is only about 2%.

17. *Homo floresiensis* is known from deposits dated to less than 100 kya (the best-preserved individual lived about 18 kya) on the island of Flores in the Indonesian archipelago. Some have claimed that these tiny hominins (there are several individuals) were merely pathological modern humans, though most paleoanthropologists today accept them as a separate species that split from our ancestors a long time ago. It doesn't really matter for this discussion though. The point is that these small-brained hominins engaged in the sorts of behaviors (making tools, hunting, possibly controlling fire) that we consider part of the *Homo* adaptive zone. And few seriously entertain excluding *floresiensis* from the human genus.

18. Ana Navarrete, of the University of St. Andrews, and her colleagues have recently shown that there is no relationship between brain size and the mass of the digestive tract across the mammals. They found also that humans actually do have high metabolic rates when considered relative to our fat-free body mass. These researchers suggest that higher-quality diets, processing foods with tools and fire, food sharing, and reduced energy costs of walking on two legs made up the difference. See Ana Navarrete, Carel van Schaik, and Karin Isler, "Energetics and the Evolution of Human Brain Size," *Nature* 480, no. 7375 (2011): 91–93.

19. There are several other bits and pieces, especially from the Omo River Valley of southern Ethiopia, that are very likely *Homo*, but can't be identified to species.

20. William McGrew, *The Cultured Chimpanzee: Reflections on Cultural Primatology* (Cambridge: Cambridge University Press, 2004).
21. I compared them with those of the Lucy species, *Australopithecus afarensis*. We have plenty of those.
22. This idea hasn't gained much traction in the discipline, and most still consider *habilis* and *rudolfensis* to be *Homo*.

CHAPTER 7: THE NEOLITHIC REVOLUTION

1. V. Gordon Childe, *New Light on the Most Ancient East: The Oriental Prelude to European Prehistory* (New York: Kegan Paul, 1934), 2.
2. Raphael Pumpelly's *My Reminiscences* (New York: Henry Holt, 1918) documents his adventures over 75 extraordinary years. It is a riveting read.
3. Raphael Pumpelly, *Explorations in Turkestan; Expedition of 1904: Prehistoric Civilizations of Anau Origins, Growth, and Influence of Environment* (Washington, DC: Carnegie Institution of Washington, 1908), xxiii.
4. The Aral and Caspian may even have merged during what geologists today call the Khvalynian transgression. Some have gone as far as to suggest that this may have inspired ancient flood myths, including Noah's biblical narrative.
5. The abbreviation "BP" stands for "before the present"—set by convention to the year 1950. I prefer this to the more culturally loaded BC, CE, or BCE. The dates given here are based on radiocarbon dating (see chapter 5, note 8), calibrated using tree-ring dating or other techniques because of natural fluctuations in the amount of ^{14}C in the atmosphere in the past. These are often abbreviated as "cal BP" or "cal yr. BP" in the literature, and contrasted with uncalibrated dates, or RCYBP, "radiocarbon years before the present." The dates of specific sites can vary by a couple of thousand years in different publications—this is why.
6. French archaeologist Jacques Cauvin called it a "revolution of symbols." He laid out the argument in *The Birth of the Gods and the Origins of Agriculture* (Cambridge: Cambridge University Press, 2000).
7. Archaeologist Brian Hayden, at Simon Fraser University, called this "aggrandizing behavior." See Brian Hayden, *The Power of Feasts:*

From Prehistory to the Present (New York: Cambridge University Press, 2014).

8. Those familiar with Thomas Malthus's *An Essay on the Principle of Population as It Affects the Future Improvement of Society* (London: printed for J. Johnson, in St. Paul's Church-Yard, 1798), which inspired Charles Darwin to develop natural selection theory, will recognize this.

9. It is important to distinguish cultivation from domestication. Cultivation is mostly sowing and managing plants, consciously tending them rather than merely gathering them where they grow wild. Cultivated plants can be of the wild type, or of a type bred specifically for human use. Domestication, on the other hand, involves genetic modification of plants or animals by humans, whether or not it's done knowingly, to select for desirable traits.

10. Some of those grains resemble domesticated ones. Wheat and barley seeds form within spikelets that attach to a stalk. The attachments of wild forms shatter at the slightest touch or breeze, so seeds are scattered over a wide area. But domesticated ones are bred so the seeds stay on the stalk and separate only with active threshing. You can tell the difference by looking at attachment scars—smooth for wild, and jagged for domesticated. That's not to say they intentionally bred these. If you gather wild wheat and barley to plant, you'll end up with more nonshattering spikelets as the shattering type scatter during collection. More grain makes it back to camp, but the plant now requires humans to spread its seeds. At Ohalo II, there were about three times the number of jagged-scarred spikelets than would be expected had the grains been recovered from a wild stand.

11. Adnan Bounni, "Campaign and Exhibition from the Euphrates in Syria," *Annals of the American Schools of Oriental Research* 44 (1977): 1–7.

12. The term "Natufian" comes from a site in Wadi an-Natuf, near Jerusalem. Archaeologists often consider these hunter-gatherers to have been forerunners of the earliest farmers because many settled down and collected wild cereal grains.

13. Richard Alley's *The Two-Mile Time Machine* (Princeton, NJ: Princeton University Press, 2000) is a great read. Alley describes

the ice sheet, how it was cored, and what scientists have discovered from it with an authority that can only come from decades of experience.

14. There is an offset in time between the air and the ice that surrounds it because it takes a while for the bubbles to close off, but paleoclimatologists control for this in their studies.

15. Historian John Brooke at Ohio State University summarizes the evidence in *Climate Change and the Course of Global History: A Rough Journey* (New York: Cambridge University Press, 2014).

16. The name "Dryas" is a homage to an Arctic-Alpine shrub, *Dryas octopetala*. It's the national flower of Iceland, with a yellow center and eight white petals. It's a tough little plant, common in desolate tundra. Early researchers found its leaves in bog and lake deposits around northern Europe at the turn of the twentieth century, and used them as a marker for cold conditions. Three Dryas stadials are recognized—the Oldest, Older, and Younger.

17. The Baaz Rockshelter in Syria's al-Majar Depression and el-Wad Terrace in Israel's Mount Carmel area are a couple of examples.

18. Patrick Mahoney, "Dental Microwear from Natufian Hunter-Gatherers and Early Neolithic Farmers: Comparisons within and between Samples," *American Journal of Physical Anthropology* 130, no. 3 (2006): 308–19; see also Patrick Mahoney, "Human Dental Microwear from Ohalo II (22,500–23,500 cal BP), Southern Levant," *American Journal of Physical Anthropology* 132, no. 4 (2007): 489–500.

19. Thomas Hobbes, *Leviathan: With Selected Variants from the Latin Edition of 1668*, ed. Edwin Curley (Indianapolis: Hackett, 1994), 76.

20. Marshall Sahlins, "Notes on the Original Affluent Society," in *Man the Hunter*, ed. Richard Lee and Irven DeVore (Chicago: Aldine, 1968), 85.

21. Clark Larsen, *Bioarchaeology: Interpreting Behavior from the Human Skeleton*, 2nd ed. (Cambridge: Cambridge University Press, 2015).

22. The Mayo Clinic has a whole section on its website titled, "Oral Health: A Window on Your Overall Health" (http://www.mayoclinic.org/healthy-lifestyle/adult-health/in-depth/dental/art-20047475).

23. It got even worse later when table sugar became common, but we'll get to that story in chapter 8.

24. Dan Lieberman at Harvard conducted an elegant study a few years back on hyraxes fed softened and toughened foods. Higher chewing strains resulted in more growth of the lower face that anchors the teeth. You can read about this in Daniel Lieberman, *The Evolution of the Human Head* (Cambridge, MA: Harvard University Press, 2011).

25. Ron Pinhasi, Vered Eshed, and Peter Shaw, "Evolutionary Changes in the Masticatory Complex following the Transition to Farming in the Southern Levant," *American Journal of Physical Anthropology* 135, no. 2 (2008): 136–48.

CHAPTER 8: VICTIMS OF OUR OWN SUCCESS

1. Adrian Kin and Błażej Błażejowski, "The Horseshoe Crab of the Genus *Limulus* Living Fossil or Stabilomorph?" *PLoS One* 9, no. 10 (October 2, 2014), doi:1:10.1371/journal.pone.0108036.

2. George Carlin, *Jammin' in New York*, directed by Rocco Urbisci (Orland Park, IL: MPI Home Video, 2006), DVD.

3. Pythagoras, quoted in Ovid's *Metamorphoses, Book XV*, 60–142, trans. Anthony Kline (University of Virginia Library), © 2000 A. S. Kline, http://ovid.lib.virginia.edu/trans/Metamorph15.htm.

4. "Animals Used for Food," PETA, last modified August 1, 2016, http://www.peta.org/issues/animals-used-for-food/.

5. Sarah Palin, *Going Rogue: An American Life* (New York: Harper, 2009), 133.

6. Marilyn Gaull, "Byron and the Dragons of Eden," in *Byron: Heritage and Legacy*, ed. Cheryl Wilson (New York: Palgrave Macmillan, 2008), 82.

7. There's also a great quote in a letter written by French philosopher Pierre Gassendi to his friend, the Flemish physician Jan Baptist van Helmont in 1629: "All animals . . . which Nature has formed to feed on flesh have their teeth long, conical, sharp, uneven, and intervals between them—of which kind are lions, tigers, wolves, dogs, cats, and others. But those who are made to subsist only on herbs and fruits have their teeth short, broad, blunt, close to one another, and

distributed in even rows. Of this sort are horses, cows, deer, sheep, goats, and some others. And further, that men have received from Nature teeth which are unlike those of the first class, and resemble those of the second." Quoted in Rod Preece's *Sins of the Flesh: A History of Ethical Vegetarian Thought* (Vancouver: University of British Columbia Press, 2008), 161.

8. Amanda Henry, Alison Brooks, and Dolores Piperno, "Microfossils in Calculus Demonstrate Consumption of Plants and Cooked Foods in Neanderthal Diets (Shanidar III, Iraq; Spy I and II, Belgium)," *Proceedings of the National Academy of Sciences of the United States of America* 108, no. 2 (2011): 44–54.

9. Woody Allen, *Sleeper*, directed by Woody Allen (Culver City, CA: MGM Studios, 1973), film.

10. S. Boyd Eaton and Melvin Konner, "Paleolithic Nutrition: A Consideration of Its Nature and Current Implications," *New England Journal of Medicine* 312, no. 5 (1985): 283–89, and S. Boyde Eaton, Marjorie Shostak, and Melvin Konner, *The Paleolithic Prescription: A Program of Diet and Exercise and a Design for Living* (New York: Harper and Row, 1988).

11. Loren Cordain and Shelley Schlender, "The History of the Paleo Movement," transcript of podcast audio, February 27, 2014, http://thepaleodiet.com/paleo-diet-podcast-history-paleo-movement/.

12. To be fair, Eaton wasn't the first to advocate a return to the diets of our ancestors. Arnold DeVries's *Primitive Man and His Food* (Chicago: Chandler, 1952) and Walter Voegtlin's *Stone Age Diet* (New York: Vantage, 1975) came years earlier, though these and other early authors did not receive the same level of attention as the works from Eaton and his colleagues.

13. Nicole Lyn Pesce, "Celeb Diets: Experts Eye the Science Behind the Hype," *New York Daily News*, July 7, 2014.

14. "Best Diets 2017," *US News & World Report*, January 4, 2017, http://health.usnews.com/best-diet/paleo-diet.

15. "Top 5 Worst Celebrity Diets to Avoid in 2015," British Dietetic Association, last modified December 8, 2014, https://www.bda.uk.com/news/view?id=39&x[0]=news/list.

16. On the positive side, there are books like *The Paleo Diet* (New York: John Wiley and Sons, 2002) by Loren Cordain, *The Paleo Solution* (Las Vegas: Victory Belt, 2010) by Robb Wolf, *Nom Nom Paleo: Food for Humans* (Kansas City, MO: Andrews McMeel, 2013) by Michelle Tam and Henry Fong, and even *Neanderthin: Eat Like a Caveman to Achieve a Lean, Strong, Healthy Body* (New York: St. Martin's Press, 1999) by Ray Audette and Troy Gilchrist. For alternative views, consider *The Low-Carb Fraud* by T. Colin Campbell and Howard Jacobson (Dallas, TX: BebBella Books, 2014), *The Gluten Lie: And Other Myths about What You Eat* (New York: Regan Arts, 2015) by Alan Levinovitz, *Diet Cults: The Surprising Fallacy at the Core of Nutrition Fads and a Guide to Healthy Eating for the Rest of Us* (New York: Pegasus Books, 2014) by Matt Fitzgerald, and *Paleofantasy: What Evolution Really Tells Us about Sex, Diet, and How We Live* (New York: W.W. Norton, 2013) by Marlene Zuk.

17. Michael Jensen, et al., "2013 AHA/ACC/TOS Guideline for the Management of Overweight and Obesity in Adults," *Journal of the American College of Cardiology* 63, no. 25 (2014): 2985–3023.

18. See William Leonard, "Food for Thought: Dietary Change Was a Driving Force in Human Evolution," *Scientific American* 287, no. 6 (2003): 106–15.

19. Zuk, *Paleofantasy*.

20. Hans Olaf Bang, Jørn Dyerberg, and Aase Brøndum Nielsen, "Plasma Lipid and Lipoprotein Pattern in Greenlandic West-Coast Eskimos," *Lancet* 1, no. 7710 (1971): 1143–45.

21. Matteo Fumagalli, et al., "Greenlandic Inuit Show Genetic Signatures of Diet and Climate Adaptation," *Science* 349, no. 6254 (2015): 1343–47.

22. See J. George Fodor, " 'Fishing' for the Origins of the 'Eskimos and Heart Disease' Story: Facts or Wishful Thinking?" *Canadian Journal of Cardiology* 30, no. 8 (2014): 864–68, for a review of problems with Bang and coauthors' original work. Indeed, several studies have failed to confirm the benefits of fish-oil supplements for heart health in at-risk patients (e.g., see The ORIGIN Trial Investigators, "n–3 Fatty Acids and Cardiovascular Outcomes in Patients with Dysglycemia," *New England Journal of Medicine* 367 (2012): 309–18, and

Evangelos Rizos, et al., "Association Between Omega-3 Fatty Acid Supplementation and Risk of Major Cardiovascular Disease Events: A Systematic Review and Meta-Analysis," *Journal of the American Medical Association* 308, no. 10 (2012): 1024–33.

23. World Health Organization, *Oral Health*, fact sheet (Geneva, Switzerland: World Health Organization Media Centre, 2012).

24. Corruccini spent most of his professional career arguing that our occlusal problems today stem from industrial-age diets. His book, *How Anthropology Informs the Orthodontic Diagnosis of Malocclusion's Causes* (Lewiston, NY: Edwin Mellon, 1999) is a great example of how we can apply an evolutionary approach to understanding and managing the effects of diet on oral health.

25. See Jerome Rose and Richard Roblee, "Origins of Dental Crowding and Malocclusions: An Anthropological Perspective," *Compendium of Continuing Education in Dentistry* 30, no. 5 (2009): 292–300.

INDEX

Page numbers in *italics* refer to figures.

orangutans, 42, 52–53, *53*, 118, 213n7
orbital dynamics, 89–92, 94, 106–107
orthodontics, disorders of, 205, *206*,
     207–208
Osborn, Henry Fairfield, 9, 11–15,
     209n2
osteoarthritis, *191*
overbite. *See* orthodontics, disorders of
Owen, Richard, 44
Ozark Plateau, of Northwest Arkansas,
     48–49

Paleo Diet, 202. *See also* Paleolithic diets
paleobiology, 13–14
Paleolithic diet, 4, 200–203
paleomagnetism, 93
paleopathology, 190–94, *191*
Palin, Sarah, *Going Rogue: An American
     Life*, 199
Paranthropus, *28*; discovery and interpre-
     tations of, 67, 69–71, *71*, *76*, 73–79,
     81–82, 85–86, 110, *111*, 135, 139,
     141, 144, 161, 167, 214n24, 218n5,
     219n2; microwear of, 111, 117–118,
     120, 125–28, 136; stable carbon iso-
     tope analysis of, 131, *132*, 133–34,
     136, 165; tooth size of, 69, 73–75,
     78, 163
Peabody, George, 8
periodontal disease, 193–94, 205
People for the Ethical Treatment of
     Animals (PETA), 199
phenology, 48–50, 212nn9, 10
photoperiod, 212n9
photosynthesis, 216n12, 218n8; path-
     ways of, 128–37
phytoliths, 116
pigs, fossil, 79, 82, 137
Pima, peoples, 207
Pithecanthropus. See *Homo erectus*
plants: C_3, 128–29, 131, 133, 135–37,
     216n12; C_4, 128–29, 131, 133–37,
     217n12; tannins in, 46; toxins in, 45
plaque bacteria, 193, 205
plate tectonics, 88–89, 95–97, 102, 106–
     107, 138

Plavcan, J. Michael, 145
pollen, from peat bog deposits, 105
population, pull and push models of, for
     the Neolithic Revolution, 174
potlatch. *See* aggrandizing behavior
Potter, Harold ("Monkey-rope"), 68
Potts, Richard, 102–105, 108, *109*, 150
precession. *See* orbital dynamics
psycho-cultural paradigm, for the Neo-
     lithic Revolution, 173–74. *See also*
     revolution of symbols
Puerco Formation, 9
pulsed climate variability hypothesis, 107
Pumpelly, Raphael, 171–173, 222n2
punctuated equilibrium, 83, 85, 215n28
Pythagoras, 199

radiocarbon dating, 222
Red Queen hypothesis, 136, 215n28
Remis, Melissa, 43–46
Robinson, John, 68–73, 142, 145–46
Roblee, Richard, 208
rodents, 17, 26
Rose, Jerome, 208
Ross, Elizabeth, 158
Rusinga Island, Kenya, 142

Sacoglottis nuts, 55, 57
Sagan, Carl, 112
Sahara Desert, of North Africa, 87–88,
     92
San Francisco Zoo, San Francisco, Cali-
     fornia, 44–46, 212n8
savanna hypothesis, definition of, 66
scanning electron microscopy (SEM),
     115–19, 121, 150
scavenging, role in human evolution,
     150–51
Schaller, George, 41–42; *Year of the
     Gorilla*, 41
Schoeninger, Margaret, 130
Schuchert, Charles, 13
Scott, Robert, 110, 123, 125
seasonality, 37, 42, 44, 49–51, 68, 73,
     75, 89–93, 116, 154–57. *See also*
     phenology